# From Tropical Root to Responsible Food

## Enhancing Sustainability in the Spice Trade

Edited by Marije Boomsma and Ellen Mangnus

Royal Tropical Institute

## colophon

KIT Publishers, PO Box 95001, 1090 HA Amsterdam, The Netherlands
www.kitpublishers.nl

© 2011 – KIT Publishers, Amsterdam

ISBN 97894 6022 148 4

This publication is jointly produced by the Royal Tropical Institute (KIT) and the Dutch Spice Trade Association (NVS).

The conference on which this publication is based was made possible with financial and organizational support from:

**IDH**
PO Box 48, 3500 AA Utrecht, the Netherlands
www.dutchsustainabletrade.com

**Royal Tropical Institute (KIT)**
PO Box 95001, 1090 HA Amsterdam, the Netherlands
www.kit.nl

**Dutch Spice Trade Association (NVS)**
Einsteinstraat 30-G NL, 1446 VG Purmerend, the Netherlands

**CREM**
Spuistraat 104D, 1012 VA Amsterdam, the Netherlands
www.crem.nl

**Cordaid**
PO Box 30919, 2500 BK The Hague, the Netherlands
www.cordaid.nl

**Both Ends**
Nieuwe Keizersgracht 45, 1018 VC Amsterdam, the Netherlands
www.bothends.nl

KIT and NVS are grateful to all the individuals who contributed to this publication: Mark Barnett, Patrick Barthelemy, Karin Bogaers, John Fagan, Michael Gravina, Jose Ruijter, Cleopa Ayo, Bharat Maskai, Samash Natu and Alex Bruijnis, Tjeerd de Vries and Bart de Steenhuijsen Piters.

*Coordination*
  Marije Boomsma and Ellen Mangnus

*Editing*
  Sarah Simpson (KIT) and Valerie Jones, Contactivity bv, Leiden
  www.contactivity.nl

*Design*
  Grafisch ontwerpbureau Agaatsz bNO, Meppel

*Printing*
  High Trade bv, Zwolle

# Table of contents

| | | |
|---|---|---|
| | Acronyms and abbreviations | 4 |
| 1 | Introduction | 7 |
| 2 | A brief history of the spice trade | 11 |
| 3 | Drivers of sustainability in the spice sector: ethics and continuity | 15 |
| 4 | Sustainability in spice production and trade | 19 |
| | 4.1 Cassia Co-op, Indonesia | 20 |
| | 4.2 Elephant Pepper Mozambique | 24 |
| | 4.3 Pacific Basin Partnership, Vietnam | 29 |
| | 4.4 Laxmi Enterprises, India | 33 |
| | 4.5 Uganda Crop Industries Limited | 36 |
| | 4.6 Discussion | 42 |
| 5 | The food and retail sector | 47 |
| | 5.1 Royal Ahold | 47 |
| | 5.2 Discussion | 51 |
| 6 | Regulations and certification | 55 |
| | 6.1 Certification for a sustainable world | 55 |
| | 6.2 The role of governments | 64 |
| | 6.3 Discussion | 70 |
| 7 | Managing relationships | 73 |
| | 7.1 Golden Food Products, Tanzania | 73 |
| | 7.2 Discussion | 79 |
| 8 | Towards sustainable trade | 81 |
| | 8.1 A short synthesis | 81 |
| | 8.2 Epiloque | 82 |
| 9 | The way forward | 83 |
| | About the authors | 85 |

# Acronyms and abbreviations

| | |
|---|---|
| ASTA | American Spice Trade Association |
| BSCI | Business Social Compliance Initiative |
| Both ENDS | Environment and Development Service (the Netherlands) |
| BRC | British Retail Consortium |
| BSE | bovine spongiform encephalopathy |
| Cordaid | Catholic Organisation for Relief and Development Aid (the Netherlands) |
| CREM | Consultancy company specializing in sustainable development |
| CSR | corporate social responsibility |
| EL&I | Ministerie van Economische Zaken, Landbouw en Innovatie / Ministry of Economic Affairs, Agriculture and Innovation (the Netherlands) |
| EOS | Earth Open Source |
| EPOPA | Export Promotion of Organic Products from Africa |
| EtO | ethylene oxide |
| EU | European Union |
| Faida MaLi | Faida Market Link Company Ltd (Tanzania) |
| GFP | Golden Food Products (Tanzania) |
| GFSI | Global Food Safety Initiative |
| GlobalGAP | Global Good Agricultural Practices |
| GMO | genetically modified organism |
| HACCP | Hazard Analysis Critical Control Point |
| IDH | Initiatief Duurzame Handel / Dutch Sustainable Trade Initiative |
| ILO | International Labour Organization |
| IMO | Institute for Market Ecology (Switzerland) |
| ISO | International Organization for Standardization |
| Juwakihuma | East Usambara Organic Farmers' Association (Tanzania) |
| KIT | Royal Tropical Institute |
| NGO | non-governmental organization |
| NVS | Dutch Spice Trade Association |
| OFI | Organic Food Ingredients Sagl (Switzerland) |
| PBP | Pacific Basin Partnership (Vietnam) |
| PSI | Private Sector Investment Programme (the Netherlands) |
| RSPO | Roundtable on Sustainable Palm Oil |
| RTRS | Round Table on Responsible Soy Association |

| | |
|---|---|
| SMEs | small and medium enterprises |
| TanCert | Tanzania Organic Certification Association |
| TOAM | Tanzania Organic Agriculture Movement |
| UCIL | Uganda Crop Industries Limited |
| USADF | United States African Development Foundation |
| VOC | Verenigde Oost-Indische Compagnie / Dutch East India Company |
| WWF | World Wildlife Fund |
| WTO | World Trade Organization |

# 1 Introduction

Spices refer to a diverse group of dried seeds, fruits, roots, bark and other plant materials that are used as seasonings or to add colour or aroma to a wide range of foods and beverages. Worldwide, the demand for spices of all kinds has increased steadily in recent years, as a result of population growth and rising incomes in India, China and other developing countries, and the changing Western preferences for exotic and spicy foods.

In 2009, the volume of spices traded on global markets reached 4 million tonnes. Most of these spices originate from developing countries in Africa, Asia and Latin America, where they are often produced on small farms by very poor farmers. The large number of small-scale farmers involved means that the spice industry can have a significant influence on the incomes of poor families.

The largest producers of spices are China and India, where the preference for spicy dishes means that most spices are used for domestic consumption. The largest overseas markets are European countries, particularly Germany, the Netherlands and the UK, which import about 23% of the volume of all spices traded, followed by the United States at about 16%.

Perhaps surprisingly, although the Netherlands is not a producer of spices, it is the largest importer and re-exporter of spices in the European Union. The Dutch involvement in the spice trade can be traced back to the 17th century, a period known as the 'Golden Age' when the Dutch were important sea traders. Since then, many trading companies and food companies that use spices in their products have been based in the Netherlands.

Despite the strong market position of companies in Europe and the United States, in recent years the competition with Asian markets has become fierce, and is expected to grow due to the rising domestic demand in many Asian countries. Western buyers generally offer better prices but they face a competitive disadvantage due to the fact that they also require stricter food safety and health guarantees. On top of that, Western retailers and consumers are demanding ever higher standards with regard to the social and environmental aspects of the food they buy. This makes it harder for producer countries to export to these markets.

Regulations and standards require good production inputs and practices, the application of quality management standards, traceability of products and approved social conditions. These are all conditions that require substantial investments in farm production and supply. Many farmers and agri-trade companies in developing countries lack the resources to invest in improving the quality of and expanding their production. They do not have sufficient private capital, nor are they able to attract foreign equity. As a consequence, farmers and spice exporting companies are abandoning export markets in favour of local and regional markets, or are switching to other crops such as cocoa or oil palm.

The growth of the artificial flavour industry is adding further pressure. Increasingly, natural spices are being replaced by synthetic substitutes, which are usually cheaper to produce and easier to source.

In order to retain their market shares, Western traders and food processors are looking for ways to secure more reliable supplies. One option is to diversify markets and to increase sourcing from continents other than Asia. Another is to invest in sustainable relations with producers. However, neither of these options is likely to provide an immediate return on investments.

Growing domestic markets in Asia, strict import requirements, sustainability trends and competition from the artificial flavour industry are threatening the spice sector. The Dutch Spice Trade Association (NVS) has taken up the challenge of helping the industry find sustainable solutions. A few frontrunner companies are leading this mission. The majority of the Dutch companies are too small to make significant investments or are unwilling to engage in risky business ventures. However, they are interested in learning from the effort to move towards a more sustainable spice trade and are seriously considering adopting new practices, once they are proven to be worthwhile.

**Sustainable spice conference**
The Royal Tropical Institute (KIT) has long been interested in the history of the spice trade, developing expertise in sustainable businesses and supply chains, and in particular the potential contribution of the spice sector to sustainable development for the poor. Thus, in October 2010, on the occasion of its 100th anniversary, KIT decided to organize a conference to consider ways to enhance sustainability in the spice trade, together with the Dutch Spice Trade Association and the Sustainable Trade Initiative, as well as Cordaid, CREM, Both ENDS and other key players involved in spice cultivation, production and trade.

The conference brought together 150 representatives of the spice industry, the Netherlands government, consultancy firms, universities and non-governmental organizations (NGOs) to discuss the issues that play a role in the sustainability of the sector, and ultimately to ensure the commitment of all stakeholders to the future development of the spice industry.

The spice industry faces a number of unique sustainability challenges and opportunities. The complexity of sourcing, the length of supply chains, the diversity of participants in each chain, both culturally and economically, and the penetration of the spice chain into virtually every branch of the food system, all contribute to the challenges. It is time to reflect on possible solutions to help the spice sector become more sustainable.

This book gives an overview of the different issues facing actors in spice chains. Chapter 2 provides a brief history of the spice trade. Then, in Chapter 3, Marianne van Keep, chair of the NVS, presents her views of sustainability. Chapters 4-7 present a series of case studies of businesses and other organizations associated with the spice sector. Each chapter concludes with a summary of the discussions that took place during the conference. In Chapter 8, KIT draws together some general conclusions on the most urgent sustainability issues and its vision of a sustainable spice trade. Chapter 9 closes with presenting the future plans of the Dutch spice sector.

# 2 A brief history of the spice trade

Even in ancient times, Asian merchants were involved in the spice trade. The Arabs first monopolized the trade routes from East to West, buying spices and herbs from Chinese or Javanese merchants and shipping them from India to Egypt. Alexandria was the trade centre from where spices entered the Greek and Roman empires. For each spice, the chain from producers to consumers was often long, and spices changed hands many times before reaching Europe. The Arabs were renowned as keen traders, doing everything to confuse buyers about the origin of the spices in order to maximize their profits.

During the Middle Ages, the Ottoman Empire took over the maritime trade routes from the Arabs, and Muslim traders dominated the spice trade for eight centuries. It was not until the Age of Discovery that Europe gained control of the spice trade. In 1499, the Portuguese navigator Vasco da Gama was first to reach India after sailing around Africa. He enforced trade agreements with Indian rulers to supply spices such as cloves, nutmeg, ginger and pepper. Also in this period, the explorer Christopher Columbus reached the Americas, and introduced new spices, including allspice and vanilla, to European consumers.

Over the next century, competition to gain control of the spice trade was fierce. Portugal, Spain, England and the Netherlands all established their own trade routes, forts and alliances and competed to gain a monopoly on as many spices as possible. This period of economic and political expansion/settlement was supported by a growing class of merchants who were prepared to invest in building the first big trading corporations. Increasingly trade became more organized.

For decades the Portuguese were the most powerful players in the spice trade. They were then overtaken by the Spanish, who gained a monopoly in 1580. Dutch merchants were also eager to engage in the trade, and to build ships to bring the spices back in large quantities. But the large capital cost of building a fleet and the threat of pirates made it too risky a venture for individual investors. So they decided to pool their resources, and in 1602 formed the Dutch East India Company (VOC). The VOC had a huge impact on the Dutch economy. For the next 200 years, most jobs in the Netherlands were related, directly or indirectly, to the VOC. Other European countries adopted similar strategies, but none was as successful as the VOC. By 1670 it was the richest company in the world, paying its shareholders an annual dividend of 40% on their investment, despite financing 50,000 employees, 30,000 fighting men and 200 ships, many of them armed. This lasted until 1780 1784, when the British took over the Dutch trading posts in India and became the main trading power.

Over time, spices were no longer scarce commodities, and monopolies slowly gave way to markets. For years the secret of the spice trade was simple: maintain demand by keeping a tight control on supplies, either by force through political power or by arrangements between merchants, although supplies were occasionally affected by weather conditions, such as droughts or heavy rainfall in the producer countries. But although the spice trade was no longer controlled, the flow of information was still imperfect and practiced traders continued to make a lot of money. The spice bazaars in Kerala (India), Ambon (Indonesia) and Rotterdam were hectic and dynamic places.

Some features of the spice trade remain unchanged today. Most spices are still grown in tropical or subtropical countries where the climate and soils are most favourable. Each chain still involves many players, and spices still change hands many times. Competition in the market is just as fierce, and may be becoming even fiercer with the introduction of synthetic flavourings as new competitors. And buyers are still looking for high quality.
The context in which spice trade operates has changed dramatically, however. Market information is now freely available, and farmers in the most remote

areas can access up-to-date prices via their mobile phones. And the transfer of spices along each chain has been reduced from months to only a few weeks. Problems with pirates, storms and scurvy may be things of the past, but have been replaced with new obstacles: quality standards and certification systems in the importing countries. The power of the buyer is no longer enough to guarantee the delivery of spices; other actors such as traders, producers and retailers are increasingly important. And, in order to ensure that the spice trade has a future, many consumers and companies have added a new condition – that the spice trade should be sustainable.

**Spices and their uses**

Spices are used as food additives to provide flavour, colour or fragrance, or as preservatives that kill or prevent the growth of harmful bacteria. Many spices are also used in medicines, cosmetics and perfumes, and some are eaten as vegetables. For example, liquorice is used as a medicine and garlic as a vegetable.

In the past spices also served other uses. In ancient Egypt cassia (or cinnamon) fetched a high price because it was essential for embalming. Anise, marjoram and cumin were used to wash the bodies of the dead. In both China and India, traditional systems of medicine were based on herbs, and are still widely used today. In Europe, spices were used mainly to preserve perishable vegetables and meats, and to make poorly preserved food edible during the long winter months. During the Middle Ages spices were used as medicines and to protect against pests and diseases.

# 3 Drivers of sustainability in the spice sector: ethics and continuity

Marianne van Keep, chair of the Dutch Spice Trade Association, and director of purchasing, Verstegen Spices & Sauces BV, Rotterdam, the Netherlands

After so many years, it is hard to remember what put sustainability on the agenda in the spice sector: was it personal conviction? A request from an NGO? Economic interests? Certainly the ethical dilemmas facing buyers in their day-to-day activities may have played a role. Nearly every decision needs to be evaluated, and economic interests are often weighed against the interests of society.

Corporate social responsibility (CSR) is becoming increasingly important in many decisions, particularly with regard to procurement. One frequently used example of CSR involves refusing to purchase or handle goods produced by child labour, defined as 'work by children under age 15 that harms their physical or mental development'. According to this definition, 218 million children worldwide are child labourers, 126 million of whom work under dangerous conditions. Many are believed to be working in the agricultural sector, which of course implies that children are also working in the spice sector. The buyers of many companies active in the spice sector, including Verstegen, travel regularly to spice-growing countries and they have not encountered child labour during their visits, as far as we know. Nevertheless, most buyers make a habit of raising the issue with their local contact, emphasizing that child labour is not acceptable. It is also mentioned in procurement contracts, and new suppliers receive a letter that explicitly refers to the issue. But we realize that much more can be done to make child labour *unnecessary*: by making sure that producers earn enough so that their children do not have to work. It's as simple as that. Or is it?

Just as important as social responsibility to the sustainability discussion is the issue of the continuity of supply. If we want to be sure of stable supplies of spices in the future, we have to ensure that the spice industry is attractive to producers and processors so that they remain in the industry. What matters most for producers is a good price.

In 2005, Verstegen launched its first regional project, processing nutmeg and mace on the north coast of Ambon, in Indonesia. Primary products are sourced locally as far as possible, and are bought at higher prices than those offered by traders. The producers are trained in how to maintain high quality and they receive a bonus when the produce they deliver meets requirements. The project has also established a processing facility that complies with international hygiene norms and employs about 60-70 labourers, mainly women. As a result, the entire community benefits. Of course, with billions of people working in agriculture worldwide, this project is but a drop in the ocean. Most farmers would like to earn more, if only to be able to pay for education for their children and healthcare for their families. But between companies like Verstegen and the producers there are collectors, traders and processors, and in some cases, speculators. So how can we ensure that more money reaches producers?

Ultimately, most companies can probably achieve greater impact if they collaborate at industry level. In 2008, the Dutch Spice Trade Association (NVS) set up a working group on socially responsible entrepreneurship. It made a conscious effort to include processors as well as traders, because, although each group has its own role and interests, when it comes to chain responsibility these different interests have to be aligned.

The members of the working group are known to many organizations active in the field of socially responsible business. For example, we are currently talking with the NGO Fairfood about how we can comply with their credo 'eat fair, beat hunger'. We have also launched a major project, together with CREM, a consultancy, and Both ENDS, to analyze a number of product–country combinations, such as ginger and turmeric from India, and pepper and nutmeg from Indonesia. In the spice sector, there is an immense diversity of products and countries of origin. This is a key difference with other industries that are collaborating on the issue of socially responsible business, such as coffee, tea and cocoa. For these three commodities, volumes are extremely large while the diversity of product types and source countries is rather low. By studying a number of spice products and producer countries, we hope to be able to focus our efforts, and apply the knowledge gained to other product–country combinations, such as ginger from Nigeria or chilli from India. Once the chain analyses have been carried out, we will have to make choices regarding which social issues to address. This could be anything, from combating poverty or child labour, to improving labour conditions, the environment or biodiversity.

Another question that needs to be addressed is whether the spice sector should invest in developing its own certification system or apply existing certification schemes. Certification systems serve to demonstrate that the producers of food products comply with certain standards, or to indicate fair trade or organic products. Since such systems are expensive to introduce and maintain, however, it is perhaps questionable whether the investments required would be better spent on, for example, community development projects or for other purposes that would benefit small farmers.

## Certification systems

A wide range of certification systems have been introduced in response to growing consumer interest in the economic, social and environmental impacts of trade. Such systems assign labels to products that comply with a defined set of standards. Some examples of certification programmes are as follows:

**Business Social Compliance Initiative (BSCI):** The BSCI system aims to improve social standards in production facilities worldwide. The BSCI, whose members include retailers, manufacturers and importers, focuses on monitoring as an ongoing process to improve social standards, and audits using standardized questionnaires and evaluation schemes. www.bsci-intl.org

**EU Ecolabel:** The European Union's Ecolabel is a voluntary scheme, established in 1992, that applies to non-food, non-agricultural products and is intended to encourage businesses to market more environmentally friendly products and services. The EU Ecolabel is part of a broader action plan on sustainable consumption and production adopted by the European Commission in 2008. ec.europa.eu/environment/ecolabel/

**Fairtrade Labelling Organizations International (FLO):** FLO is a body that assigns its 'fairtrade' logo to products that comply with a number of social and environmental standards. Examples of such standards are that farmers are paid a fair price, are protected and thus can build a more sustainable future, and that national ecosystems are not degraded. www.fairtrade.net

**Global Good Agricultural Practice (GlobalGAP)** is a private sector body that sets voluntary standards for the certification of agricultural products around the world. The aim is to establish one standard of Good Agricultural Practice (GAP), with different product applications that can be fitted to the whole of global agriculture. Certification is carried out by more than 100 independent accredited bodies. www.globalgap.org

**International Federation of Organic Agricultural Movement (IFOAM):** IFOAM's standards are regarded as the benchmarks for organic agriculture worldwide. These standards focus on water and soil conservation and the protection of biodiversity, but also prohibit the clearance primary ecosystems, the use of genetically modified products and of course of agrochemicals. www.ifoam.org

**ProTerra:** Cert ID's ProTerra certification programme provides impartial, third-party, verification that food and feedstuffs have been produced sustainably, in a manner that is both socially and environmentally responsible. ProTerra achieves this in a way that works for industry. The Earth Open Source standard is based on ProTerra. www.cert-id.com/proterra

**Rainforest Alliance:** The Alliance's certification and verification standards combine the aim of preserving forest biodiversity with efforts to transform land-use practices, business practices and consumer behaviour, and to improve the living conditions of producers and workers in developing countries. www.rainforest-alliance.org

**UTZ Certified:** UTZ Certified has introduced a code of conduct with social and environmental standards for commodity sectors such as palm oil, cocoa and coffee. UTZ Certified has established an online 'track-and-trace' monitoring system that enables purchasers to follow the movements of produce along each chain. www.utzcertified.org

In all of these certification systems, companies are required to undergo social audits, processes that enable them to assess and demonstrate their social and environmental benefits, and indicate where improvements are needed.

Meanwhile, the working group has established close contacts with the Sustainable Trade Initiative (Initiatief Duurzame Handel, IDH) and the Royal Tropical Institute (KIT), organizations with extensive expertise in the field of corporate social responsibility and sustainability in the agricultural sector. We are also taking further steps together with IDH and KIT and the focus seems to be turning towards biodiversity and socio-economic issues.

In 2010, the international year of biodiversity, growing spices has much to offer in that regard. Spices are generally 'non-timber forest products', meaning they are grown on small parcels of land in combination with other crops. When producers earn enough income from spices, they will be less inclined to switch to other, more intensive crops like oil palm or rubber. In addition, 'spice gardens' often function as buffer areas between farms and forests. And because those involved in spice production are overwhelmingly small-scale farmers, the industry supports poverty alleviation and income generation, which in turn will reduce the need for child labour.

The statutes of the Dutch Spice Trade Association, established in 1916, state that its aim is 'to promote the spice trade, the spice industry and their interests, to protect the trade and the industry against unfair and unlawful actions and to counter misuse of the trade and industry'. Socially responsible entrepreneurship fits both this aim and our desire to ensure the future of the spice sector. Major challenges remain, and cooperation and joint action is perhaps the greatest challenge of all. Together with our partners mentioned above, we can prove that it is possible to meet all of these challenges.

# 4 Sustainability in spice production and trade

Most spices are produced by poor, small-scale farmers who grow them as secondary crops alongside food crops such as maize and bananas. These farmers mostly sell their spices on the spot market, or to traders who buy them directly at the farm gate, or the producers themselves deliver them to a collection point. Often, farmers sell to the same buyers year after year, although in principle they are free to sell to anyone. For a farmer to commit to a buyer, he or she needs above all guaranteed and competitive prices for their produce. Supply contracts may work and help smallholders to access markets, buy they do not provide guarantees to the buyer. People who are struggling to improve their livelihoods will always consider better offers. If prices are not rewarding the farmer might be inclined to shift to producing higher-value crops. In Indonesia, for example, many farmers have recently switched from spices to cocoa production.

A spice exporter/trader has to deal with farmers but also with buyers. He or she is the bridge between spice production and the market. On the one hand, the exporter/trader has to deal with hundreds of small-scale farmers, usually through intermediate traders, who all want a good price for their produce. On the other hand, exporters need to deliver products that comply with quality requirements and, increasingly, with social and environmental standards in volatile markets. This requires costly quality management systems and training for farmers. Exporters/traders are the interface between the world of smallholders and that of consumers. In the case of international trade, they have not only to establish a bridge between poor and rich people, but also between cultures. This requires sound business skills, a thorough understanding of the diverse actors in the value chain and excellent communication skills.

This chapter presents five case studies that examine the opportunities and challenges faced by farmers, businesses and exporters in Indonesia, Mozambique, Vietnam, India and Uganda, and how they are dealing with the issue of sustainable production.

## 4.1 Cassia Co-op, Indonesia

*Patrick Barthelemy, Cassia Co-op, West Sumatra, Indonesia*

Cassia Co-op was established in 2010 and is Europe's first registered cassia (cinnamon) cooperative. It is owned by its members, who include employees and market players in the West. Its vision is to be the bridge between cassia farmers and end-users, a marketing channel for the farmers and a direct link to the source for end-users.

Cassia Co-op's mission is to promote the vertical integration of the cinnamon supply chain in order to guarantee quality, secure availability, add value at origin and in the process benefit its members. Quality control starts with the selection of the raw material. Farmers are aware of the quality required to meet Cassia Co-op's high standards of quality and cleanliness. An estimated 60% of the total demand for spices is from industrial food manufacturers that require good specifications, reliable suppliers and quality assurance procedures.

Cassia Co-op aims to be known as a dynamic, ethical company whose core values of fairness, accountability and transparency underpin its relationships with farmers, customers, associates and partners. Ethical business is not just about paying a price premium, but primarily about changing the relationship between farmers and consumers into one that is more transparent, more equitable and more efficient.

### What is cassia?

Cassia – or Indonesian cinnamon, as it is known to Western consumers – is the dried inner bark of tropical evergreen tree *Cinnamomum burmannii blume*. It is produced in agroforestry systems by farmers near Kerinci, in the province of West Sumatra, Indonesia. The farmers sell the cassia to a handful of exporters located in Padang at the coast, eight hours away.

Natural cassia is used as an ingredient in various foods and beverages, including bakery and confectionery products, cereals and teas. Valued for its flavour and fragrance, cassia is also used as a seasoning, in potpourri and decorative items, and in pharmaceuticals and 'nutraceuticals', or food-based dietary supplements.

The market for cassia is strong and growing for a number of reasons, in particular the increasing demand for spicy and flavoured foods, and for 'natural' and organic products. Cinnamon also offers some newly discovered health benefits, and is being used in the fight against diabetes and cholesterol, among others. The economic recession is another factor; history has shown that during recessions people actually tend to consume more food rather than less.

Cassia is an established natural food flavour that is difficult to imitate. The earliest product to imitate cassia flavour was synthetic cinnamic aldehyde. This is a crude imitation, however, that is no substitute for the flavour of natural cassia.

### Sustainability issues in the cassia supply chain

Until 1997, Kerinci farmers regarded cassia as their key crop and a source of guaranteed income. In fact, cinnamon trees functioned as a kind of 'bank' for farmers, which they could cut down and sell in the event of an emergency or when they needed money to pay for major items such as school fees.

But the market for cassia is volatile, and the prices paid to farmers can vary enormously. In 1997, the average price for raw cassia bark was €2.60/kg, but today it is less than €0.60/kg. Even after 13 years of inflation and the growing cost of living in Indonesia, cinnamon prices are still 75% lower than they were in 1997. On top of that, cassia farmers are isolated from market information by a chain of local dealers and collectors. They are 'price takers' in the chain and have no leverage to negotiate.

After years of hope that the cinnamon market would become profitable again, farmers have come to the harsh conclusion that planting cassia trees and waiting 15 years to collect the bark at €0.60/kg is no longer a profitable option. They have cut their losses and have started to chop down the trees, sell the bark and grow more profitable crops, such as chilli, potatoes, cocoa and rubber. This can have serious environmental impacts, due to the loss of biodiversity and the increased use of fertilizers and pesticides needed to grow other crops.

### Sustainable solutions

There is a global trend in the food industry to want to get closer to their sources of raw materials. Large customers are no longer looking for 'one-stop' vendors, but prefer to buy from specialized suppliers that provide value and expertise.

For liability and safety reasons, customers expect suppliers to be accountable for the quality and traceability of their products. In addition to meeting those growing concerns, cutting out the middlemen makes sense from an accounting point of view. It translates into better prices for customers (end users) and higher profit for the company, some of which can be passed on to the farmers (added-value coupon). This is what Cassia Co-op is aiming for.

The strength of Cassia Co-op is the combination of its presence in the area where the cassia is grown as well as in the major consumer markets. The traditional supply chain for cassia includes many actors, while Cassia Co-op aims to simplify the supply chain and to become the bridge between farmers and end users (see figure).

> **Traditional cassia supply chain:**
> Farmer ⇨ local ⇨ collector ⇨ local trader ⇨ exporter ⇨ broker ⇨ importer ⇨ processor ⇨ distributor ⇨ end user
>
> **Cassia Co-op supply chain:**
> Farmer ⇨ Cassia Co-op ⇨ end user

From the company's point of view, having farmers involved in the programme makes economic sense in terms of traceability, information, availability and quality control. It enables the implementation of organic and Fair Trade certification. Eliminating the middlemen means that the company is able to increase its gross profit margin, which in turn leaves more room for profit sharing and investments, such as in added-value production.

From the farmers' perspective, participating in the programme brings financial benefits and information about the changing demands of end users. Farmers value this transparency, which contributes to building trust between them and their marketing company. A processing plant at the source which brings added value to their crop is an indirect financial benefit and a source of pride and empowerment.

**Cassia Co-op added value scheme**
Cassia Co-op now has more than 400 farmers who have agreed to participate in the cooperative. Only bark supplied by these 'listed farmers' can be sold to the collection facility. Cassia Co-op keeps track of each farmer's output and can guarantee traceability of the raw material all the way back to the field.

The company has introduced a scheme to ensure that the farmers benefit from the value added to the product by cleaning, processing, packaging, shipping, marketing, customer service, etc. Each time a farmer delivers at least 500 kg of raw cassia, he receives payment in cash, plus an added-value coupon equal to 3% of the total revenue generated by the same net weight of product, processed, packaged and sold by Cassia Co-op to the final customer (see box). That 3% of the selling price represents an increase of about 6% on the original price at which he sold his raw cassia to Cassia Co-op.

The scheme benefits both the farmers, who earn higher incomes, and the company in the form of from increased farmer loyalty and more reliable supplies.

> **Cassia Co-op's value-added coupons**
>
> On 4 June, a farmer sells 1000 kg of raw cassia bark to Cassia Co-op, for which he receives a cash payment plus an added-value coupon that he can cash in four months later, on 4 October.
>
> If the local market price of raw cassia on 4 June is 6864 rupiah (IDR) per kilo (about ¤0.06/kg), then the farmer will be paid IDR 6,864,000 (about ¤600) for his 1000 kg.
>
> Over the next few months, Cassia Co-op cleans, processes and packages the cassia, adding value to it at each step, and then sells it. On 4 October the farmer can cash in his added-value coupon. Each month, Cassia Co-op posts the average selling price of processed cassia in the previous month. If the selling price in September was IDR 13,728/kg, then the coupon will be worth 3% of that, i.e. IDR 411.84/kg, or IDR 411,840 (¤36).
>
> Thus, for his 1000 kg of raw cassia, the farmer will have received IDR 6,864,000 plus an additional bonus of IDR 411,840 representing his share of the value added.

Because of the proximity of Cassia Co-op's collection facility to the production areas, the farmers can and must visit the cooperative to cash in or collect their coupons in person. This ensures that the added-value bonus goes directly them.

## Future challenges

To ensure continuous supplies of cassia in world markets, prices need to be maintained at reasonable levels. Cassia Co-op, however, does not believe that charity will solve existing problems. Sustainability issues can only be solved by creating a fair and efficient market environment. The supply chain structure must be competitive and logical. If any link in the chain is being exploited, then it is only a matter of time before that link will break.

Economically, the free market must find its own balance. Companies that offer a better distribution of the profits will naturally better secure their supplies, and will secure their position with end users, which in turn will benefit farmers.

Corporate social responsibility programmes are a practical way for companies to get involved at the grassroots level. Companies that are interested in strengthening and securing their supplies have an interest in being involved with farmers to build trust and a long-term supply. Cassia Co-op's social programme, for instance, has set up a soccer academy to help local schools develop their own soccer programmes. These give local kids the opportunity to learn the 'beautiful game' in an organized and supervised environment and have fun in the process. It is a social platform for kids, parents and volunteers from which other initiatives and activities are able to grow.

Finally, at the environmental level it makes sense for farmers to diversify the range of crops they grow in order to avoid becoming dependent on a single product. Cassia Co-op encourages growers to continue planting cassia trees, but also introduce other crops compatible with the forest environment, such as coffee or cocoa.

For more information, visit www.cassia.coop

## 4.2 Elephant Pepper Mozambique

Michael Gravina, Elephant Pepper, Mozambique

Elephant Pepper was started in 2003 when I and my new partner, Loki Osborne, decided to bring together our interests in elephant conservation, development and business.

The main reason for setting up the business was to create trade opportunities for marginalized communities where ecological conflicts – in this case between elephant and human populations – were undermining food security and ecological sustainability. But the Elephant Pepper buyers also benefit, as they are able to work with farmers that previously could not be reached, and are able to diversify their supply chains as part of their risk mitigation strategies. In addition, the production of chilli peppers in Mexico and other countries in South America is decreasing as these economies start to modernize and people are no longer interested in farming. Africa is in a position to take advantage of this production gap.

For many rural communities across Southern Africa, the crops grown by smallholder farmers are their main source of food. Unfortunately, they are also irresistible to elephants, which eat the produce and trample everything in their way. The conflicts between farmers trying to protect their farms and the elephants have led to heated battles, with death and injury on both sides. A solution was urgently needed, one that would be environmentally sustainable and friendly to animals. The answer was found in chilli pepper. Chilli-based products have been found to repel elephants from farms significantly faster than traditional methods, and from this the Elephant Pepper company was born.

Elephant Pepper began to educate farmers and many partner organizations on how to use the chilli-based deterrents to keep crops and people safe from elephants, and vice versa. More than 3000 farmers have now been trained and are cultivating fields of chillies around food crops. The next phase of the business was to seek solutions that would provide opportunities for the farmers to improve their livelihoods.

This resulted in the launch of the Elephant Pepper brand, which now accounts for 10% of the total business. The focus of this operation was the manufacture and sale of chilli sauces and other branded products using the chillies grown by the same small-scale farmers. In order to ensure future economic stability for farmers and safety for elephants, 10% of the profits from the spice products are returned to Elephant Pepper projects related to pepper production and educating farmers on how to coexist with wildlife.

Elephant Pepper has deployed this model successfully across Southern Africa, and provides incomes for small-scale farmers in Botswana (Okavango delta), Kenya (Laikipia), Namibia (Caprivi Strip game park), Zambia (Livingstone, Upper Zambezi, South Luangwa National Park) and Zimbabwe (Zambezi Valley).

Each Elephant Pepper farming operation produces and processes the chillies and exports chilli mash under an exclusive contract to the American McIlhenny Company as one of only seven approved suppliers for its Tabasco® brand products. Over the past six years, Elephant Pepper has exported 800 tonnes of Tabasco chilli mash to the McIlhenny Company. During the 2009 season, 300 tonnes of mash were exported from Swaziland, Zambia and Zimbabwe.

**Piri Piri Elefante Mozambique**
In 2010, Elephant Pepper extended its operations to southern Mozambique, where it will develop production to increase the volume of Tabasco chilli mash for export to the McIlhenny Company. Piri Piri Elefante will capitalize on Mozambique's optimal growing conditions, abundant and affordable labour, and the proximity to the increasingly efficient port of Maputo, to maximize yields and profitability.

The business model applied in Mozambique will be similar to that used in Zambia. This consists of owner-operated core chilli operations with production and processing facilities, and a network of commercial and smallholder farmers. The target grower base will be 3000 small-scale farmers. This business model is attractive as the core operations guarantee the McIlhenny Company a minimum volume and a quality standard, while the network of outgrowers will allow Piri Piri Elefante to increase the overall production volumes and ultimately turnover without significantly increasing its costs. The outgrowers benefit from the guaranteed market and price, which provide clear incentives for them to invest in improving their own capabilities and achieving the highest yields possible.

For Piri Piri Elefante to work, the farmers who are very poor, living in remote and often elephant range areas, need some form of support, such as training and coaching and education. Unlike in Europe, the United States and Australia, the Mozambique government provides no support or protection for farmers, so Piri Piri Elefante requested and is now receiving additional funding from the Netherlands government's Private Sector Investment programme (PSI) to strengthen the farmers' capacities. These opportunities are to anyone within the operational areas who is willing to grow chillies.

**Chillies as a pro-poor crop**
The agronomic characteristics of chilli make it *well suited for smallholders* in countries such as Mozambique, creating employment and promoting pro-poor agricultural development. Chillies can offer smallholders a valuable yield from just a small plot.

Chillies are *relatively easy to grow*. They require limited inputs, are reasonably drought-resistant, and are one of the few crop plants that can tolerate sandy soils. Chillies can be harvested throughout the year, although the yields may be higher during moderate seasons of fall and spring. Only when temperatures fall below 12°C do the chillies stop maturing.

Planting and harvesting chillies is *labour-intensive*, requiring many hands. On a commercial farm at least 50 people are needed to transplant seedlings for a 5 ha plot, while approximately 100 pickers are required for about six months to harvest

that same plot. Even smallholder farmers who plant 0.5 ha will need to hire pickers to harvest their crop.

*Wild animals, particularly elephants, hate chillies.* Most animals that would otherwise eat crops will avoid plots of chilli. Thus growing chillies allows farmers to grow additional crops on their plots that will provide extra sources of income, while limiting the possibility that food or cash crops will be eaten or destroyed by wildlife.

*The market for chillies is large and growing.* Opportunities for supplying chilli go beyond Piri Piri Elefante and its partners. The global demand for dried chillies (for spice mixes or ingredients) and fresh chillies (for sauces) is growing at a significant pace.

**Development impact**

The Piri Piri Elefante initiative will have a number of indirect economic impacts on chilli-growing communities and on Mozambique's agricultural development.

*Demonstration effect:* Despite the country's large agricultural potential, investments have been slow to materialize. A successful demonstration project is likely to encourage other investors to take advantage of this potential.

*Skills transfer:* The Piri Piri Elefante outgrower network will help farmers to develop transferable agricultural production and farm management skills that will enable them to grow additional high-value horticultural or agricultural products for domestic, regional or global markets. The company is investing in new production capacity, and providing training in basic agricultural practices that can be applied to other types of crops.

*Empowerment of women:* The project has had a positive impact on women's labour participation: 90% of the pickers are women. In many instances, the women employed on the chilli plots are the sole providers of income for their families, especially in areas where male mortality rates are high due to HIV/Aids, accidents or illness. Often these women have had no formal education and are under-qualified for many intermediate positions such as domestic staff. Wherever possible, the project promotes women to management positions because of their reliability and general work attitude. Some important positive impacts are already visible: women are now in a position to make decisions on what to grow and how to grow it, and how to use the money they earn.

*Human capital development:* Based on its policy of developing local human capital, Elephant Pepper hires and trains local workers rather than bring in external experts. This practice guarantees that technical knowledge stays within the region, helping to build future generations of farmers, and also demonstrates the project's commitment to the community.

*Income generation:* One of the most important impacts of the Elephant Pepper business model is that it provided farmers with disposable incomes. Many farmers have confirmed, for example, that the project has significantly improved their living standards, enabling them to pay for groceries, medical treatment and school fees. People are now able to save, and their increased buying power has benefited many local traders and village shops.

*Raising environmental awareness:* Farmers now have a better understanding of the environment in which they live, and thus are better able to protect it or use it in sustainable ways.

*Market structure:* The Elephant Pepper model and its development efforts will gradually change the current market structure – in which farmers bear most of the risk to grow crops that they might not be able to sell at reasonable prices – to a structure where farmers become more market-oriented, and gain increasing negotiation power to guarantee better base sale prices. Elephant Pepper guarantees to buy 100% of the chillies available, but farmers are free to sell to other buyers. Elephant Pepper completes for this business in a market-driven environment. Farmers attract the attention of other buyers and NGOs that provide training in building market linkages. Elephant Pepper does not discourage this interaction.

## Trade, not aid

Farmers in Africa face huge challenges in their lives and work. In some areas, more than 40% of the population are infected with HIV/Aids and life expectancy is low (just 35 years in the Livingstone region of Zambia). Most people do not have food security, housing or electricity. Children lack formal education, girls are often married young, and levels of domestic violence are high. Many children are being raised by their grandmothers because their parents have died from HIV/Aids-related illnesses.

Economic opportunities that tap into the entrepreneurial spirit of the world's poor are essential for ending poverty. Poor people want opportunities and access to markets. Traders and buyers are key to this. Trade, not aid, is the Elephant Pepper solution. We think this is a perfect fit with our buyers and a win-win situation for all concerned. Buyers and traders can help by investing in a supply chain in a developing country: it's good for business and essential for development.

For more information, visit www.elephantpepper.com

## 4.3 Pacific Basin Partnership, Vietnam

*Mark Barnett, Pacific Basin Partnership, Inc., Hanoi, Vietnam*

The Pacific Basin Partnership (PBP) is a manufacturer and exporter of spices in Vietnam. Established in 1992, it began with a long-term cooperation contract with Vietnam's Ministry of Trade to provide new markets for cassia (cinnamon) and other spices.

PBP sells conventional and organic spices. It imports supplies from third countries for reprocessing, and sterilizes, grinds and blends spices to customer order. The company follows American Spice Trade Association (ASTA) standards and HACCP[1] for safe, high-quality products. PBP has four field offices in rural areas as well as a factory, Son Ha Spice and Flavorings Ltd, which has 12,000 m2 under cover and four modern process lines. Raw spices are brought to the factory for drying, sorting, processing and shipment before being exported to industrial users and traders.

PBP employs 300 people, of which 75% are women. The company provides for its employees inside and outside of the workplace and makes a significant contribution to improving the lives of their families. It offers good labour conditions, including training, education for children and gender equality. Workers receive health insurance coverage from the start of their employment; light duties during pregnancy; maternity leave in excess of government standards; and at least two full meals a day for those who lodge at the factory. PBP also provides for disabled workers. Children work on the farms alongside their parents, while high-school students do piecework after school and get a hot meal.

Each year, PBP exports about 6000 tonnes of spices, including pepper, cassia and star anise. All PBP's export spices are processed in its factory. PBP is the market leader in cassia processing. It also processes hibiscus, *lo han kuo* (known as the 'magic fruit'), and green and black teas.

The company's goals are defined by consensus among management under the directors' guidance. PBP is the only spice exporter in Vietnam whose business

---

[1] HACCP – Hazard Analysis Critical Control Point – is a systematic preventive approach used in the food industry to identify potential food safety hazards, so that actions can be taken to reduce or eliminate the risk of the hazards being realized.

model puts principles before profits. PBP is committed to sustainable production and fair practice rather than 'fair trade' alone.

**Supply chain**
PBP sources from about 10,000 farmers in several areas and product groups. Pepper, cassia, star anise, hibiscus and Chinese products come from areas with very different social organizations and ethnic groups. Most pepper producers, for example, are smallholders, producing 300–1000 kg a year, who deal with a network of collectors, dealers and wholesalers. They are situated 500 850 km from the factory. PBP staff are stationed nearby and buy larger quantities from dealers who have warehouses and business licences.

The producers of cassia and star anise are members of 'hill tribe' ethnic groups who have larger land holdings. The purchasing system in these areas is more transparent, but not yet efficient. Roads are primitive and competition from nearby Chinese buyers can upset markets.

**Sustainability issues at farm level**
*Socio-economic issues.* Farmers in the more remote production areas tend to produce low value-added goods. Thus there are several processing stages from the farm to the factory, and the actors at each stage expect to make a profit. Many of these are also price speculators or stock holders. As a result, the value share is not fairly apportioned to the farmers. Spices might be produced at the farm for €1.50 per kilo, for example, but they are sold in European markets for €3 per 50-gram package, a mark-up of 4000%.

Another issue is the lack of sufficient working capital in the countryside. Most farm families are unable to save cash or build up a financial safety net, and thus do not have resources to fall back on in times of need. The failure of a crop can lead to a loss of social position and often a generation-long slide towards ever-deeper poverty.

*International standards.* The push for 'safe' products in response to the demands of consumers in Europe and North America has often come at the expense of poor farmers in developing countries. Consumers of organic products, for example, require that levels of pesticide residues are low or non-existent, and that farming practices are safe both for consumers and for the land. To meet these demands would mean eliminating chemicals at farm level. But in reality, tropical environments are home to many insects and vector-borne diseases that present health hazards both to farmers and their crops. Farmers therefore need to have the option of treating their crops with appropriate chemicals to protect both their produce and their own health.

*Subsidies and support projects.* External farm support projects, such as planting subsidies and other schemes, are often ineffective or unsustainable. PBP was once offered an unsolicited grant from a foundation in the United States to produce organic pepper. However, organic certification requires three years of 'clean' production. In a village with about 100 households, this would have been difficult to supervise. Moreover, it would have been nearly impossible to ask the farmers to abstain from all non-organic practices. Confirmation and supervision would have been difficult, and the chances of success small. Organic production requires a level of transparency among grantors and markets that could not have been met. After taking all these factors into account, PBP decided to turn down the grant.

*Diversification.* In order to improve conditions at the farm level, PBP makes direct investments in annual crops such as hibiscus, which can grow in areas with poor soil. PBP then buys the hibiscus at above-market rates (with the support of its primary customer) for processing and export. PBP has also invested in ginger and turmeric production, with mixed results. When dried, these crops can be sold in the open market, so a forward contract with prepayment does not guarantee supply.

## Sustainability issues at company level

*Market behaviour.* Local exporters, and some foreign buyers, do not take into account the farm environment or product purity in their purchasing decisions. They rely on simple tests, not on inspection and certification. This encourages intermediate dealers to use chemicals, preservatives and dyes that contaminate the supplies of spices. This in turn affects legitimate players and threatens the viability of companies as well as the markets. When the regulatory hammer falls, it will be on companies like PBP, not on the unregistered exporters and dealers.

*Prices.* Prices are determined by world markets. There are no long-term contracts with end users. That means that if another country is able to produce more cheaply during some seasons, demand for PBP's products disappears.

*Administration costs.* There is a substantial government presence in the supply chain, and value-added tax (VAT) of 5% is collected even on farm products destined for export, although it will be returned some 45–90 days later. This creates both paperwork and a financial burden for PBP. At Son Ha, the company pays about US$70,000 (€52,000) each month in VAT, or US$200,000 (€150,000) per quarter. The company bears the costs of administering all this, paying the taxes and then has to pay a percentage to the tax office to retrieve the taxes paid.

Foreign aid offered to farmers, meanwhile, is controlled by government organizations and 30% is deducted in administration fees.

## Sustainability issues at market level

*International standards.* As mentioned above, it is often difficult for farmers in developing countries to meet consumer requirements with regard to food safety and quality. For example, the US government requires that all imported food products are sterilized, but this process is harmful to the environment and produces a lot of carbon emissions. Many companies use ethylene oxide (EtO),[2] which is banned in the European Union, or irradiation. The alternative to these techniques – bacterial contamination – is not acceptable to consumers in any country.

*Market pricing.* The value of sustainability is not acknowledged by market players. For example, black pepper is the primary product in the global spice trade (other spices are marginal both to producers and consumers), but it is also the most price-sensitive: downward pressure on prices comes from end users such as supermarkets and meat processors. Consumers may be willing to pay a premium for 'sustainably produced', organically grown or equitably traded products, but the buyers at large corporations are the ones buying the bulk of pepper produced, not individual consumers.

## Conclusions

A farmer's life is the land. Sustainable production is therefore critical to farmers and their families. But in order to farm sustainably, farmers need support and a favourable environment, including stable markets paying fair prices, good information on plant protection chemicals and safe supplies, and support and guidance from other players in the value chain.

PBP believes, and acts, on the idea that it can change the conditions of farmers by giving a percentage of its gross sales revenue directly to farm and village support in healthcare, education and environmental projects.

The company hopes to become a better world citizen by making its own commitments to sustainability for its suppliers and for their children.

For more information, visit www.pbpspice.com

---

[2] Ethylene oxide (EtO) is used to sterilize products that can not be subjected conventional high-temperature steam sterilization. EtO gas infiltrates products and packaging to kill micro-organisms that remain after production or packaging processes.

## 4.4 Laxmi Enterprises, India

*Bharat Maskai, Laxmi Enterprises, Mumbai, India*

Laxmi Enterprises Group is an exporter of spices from India, supplying countries on all five continents, including Germany, Japan, the Netherlands, New Zealand, South Africa, South Korea, the United Kingdom and the United States. Laxmi has four processing plants in India, which employ a total of about 150 people. It has in-house facilities for storage (cold and ambient), cleaning, steam sterilization, grinding and packaging. The company has laboratories in Bombay and in the production areas of Vijaywada in the state Andhra Pradesh. The group works with about 750 listed farmers.

In 2007, in response to customer requirements, Laxmi decided to market spices produced without use of pesticides. This required significant investments in the activities of Laxmi's supply chain.

### Sustainability at field level operations

Having been involved in spice processing for over 15 years, Laxmi Enterprises started to realize that the spice trade is not just about segregating, sterilizing, grinding and packing spices in the factory, but starts at the production level, in the field.

In India, the Green Revolution of the 1980s saw the introduction of chemical fertilizers and pesticides on a large scale, which mostly benefitted large farmers in the north of the country. But as the price of chemicals increased over the years, more and more farmers become trapped in debt and were unable to reinvest in their businesses.

Profitability in agriculture declined and many smallholders started to drop out of the sector. The suicide rate among farmers rose dramatically. Many farmers who continued spraying pesticides started to suffer from chronic and lethal illnesses as a result. Meanwhile, Western and Japanese consumers in particular began to demand healthier and safer products.

An alternative approach to managing agriculture in more sustainable ways was clearly needed. A movement began to replace the use of chemicals with a combination of physical and biological measures, including eco-friendly bio-pesticides and agronomic soil fertility improvements, and to manage agriculture without the use of pesticides. Today, it is the ultimate goal of many farmers to receive organic certification.

Is agriculture still a profession that can sustain small-scale farming families? Sustainable agriculture requires farmers to be trained and to understand the quality requirements of the end markets, such as the European Union's requirements for traceability and pesticide residues. Farmers also need to earn enough to be able to continue growing spices year after year. Switching to organic production therefore needs to be achieved without any reduction in productivity and farm yields.

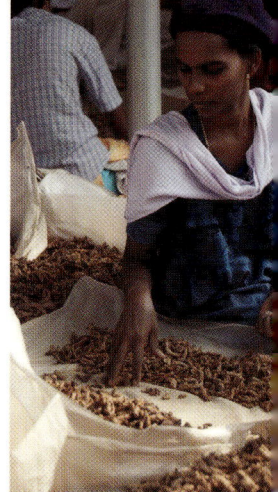

### Sustainability at trader/supply level operations

Laxmi buys from intermediate traders, in part because different varieties of spices grow in different regions of India, increasing transportation costs, and in part because in some remote areas farmers are hard to reach. When visiting traders, Laxmi emphasizes the concept of quality, and sometimes, as part of the quality assurance programme, supplies additional equipment such as instant moisture meters.

It is essential to educate traders that quality control should begin as soon as they purchase the product. Quality also needs to be managed in the processing activities of traders, such as sorting, cleaning, segregation and packing.

### Sustainability at processing level operations

Laxmi's processing plants are equipped with high-standard equipment that cater to all customer requirements. All of Laxmi's facilities are ISO 22000 and BRC A grade accredited,[3] the company is Halal registered and Kosher certified, and its laboratory is ISO 17025 2005 accredited.

Processes are controlled by means of HACCP requirements. An in-house auditor who is certified on global standards for food safety forms part of the management. This makes it possible to check quality at every stage of the production cycle with the aim to be 'first time right'. All tests required by customers are done several times before goods are shipped overseas.

Laxmi operates as a big family. Providing a safe working environment for employees is our highest priority. Laxmi provides a good shopfloor environment and interacts regularly with the staff to ensure high levels of satisfaction and motivation.

### Sustainability at the market level

At the market level, Laxmi has noted that as India's economy has grown over the last few years, customers have become more concerned about sustainability

---

3   The accreditation schemes of the International Organization for Standardization and the British Retail Consortium.

issues. They are more aware of the social and environmental aspects of the products they buy. Their requirements are passed on to suppliers, such as Laxmi. Sometimes, when changes in EU import regulations are introduced, it can be quite a challenge to meet all the new requirements. On the other hand, Laxmi sees meeting EU market standards as an opportunity to remain competitive.

## Conclusions

The transition to organic agriculture has not come easily. Laxmi encountered strong resistance from farmers and traders when they were asked to change their traditional practices. Customers, on the other hand, strongly resisted higher prices.

One problem, for instance, is to translate the high technical requirements and specifications to practical field level guidelines that the farmers can understand. Farmers also needed to become aware that that quality today is no longer a luxury, but a necessity, and that spices are food, and supplying good food maintains the health of everyone.

Despite the difficulties, Laxmi Enterprises has brought about some revolutionary changes in the production of chilli and turmeric, in terms of controlling levels of aflatoxins and other forms of microbiological contamination, and reducing the use of pesticides. In fact, it has caused a shift in thinking among farmers, away from conventional agriculture using subsidized inputs, to a system that is more eco-friendly and involves communities. As a result, over the last five years, more than 750 farmers have become involved in the new system. Since these farmers produce for European and Japanese markets, they receive 33% higher premiums than for conventional produce and this supports incomes and secure livelihoods.

The investments in the long-term sustainability of the business are being funded by the company and external parties. Some steps were funded by the government and some fixed costs were passed on to customers. Overall, Laxmi could not have achieved its ambitions without the help, financial or otherwise, of its scientific team, local government organizations, local NGO support groups and community-managed farmers' groups, as well as guidance from customers on specific product requirements.

Laxmi's business model aims to provide sustainable incomes for farmers and a sound quality system to ensure consistent supplies of quality spices to customers. The biggest challenge is to maintain our ability to pay farmers reasonable prices that cover their rising production costs. As the value of land continues to rise, so too the prices for most crops will increase. For Laxmi, it will be a challenge to decide on a fair price for any agricultural commodity.

For more information, visit www.sbslaxmi.com

## 4.5 Uganda Crop Industries Limited

Samash A. Nathu, Director, Uganda Crop Industries Limited,
Kampala, Uganda

Uganda Crop Industries Limited (UCIL), a family-owned company founded in 1993, is a diversified agricultural enterprise. The company has a long-term commitment to developing sound, sustainable and commercially viable agriculture businesses in Uganda, in partnership with small farmers, local and international public bodies, other private Ugandan enterprises, and international markets.

At the heart of the company is the 200 ha Sezibwa Estate in central Mukono district of Uganda. At the estate, the company grows sugar cane, vanilla, cardamom and timber. About 25 30% of the produce is certified organic, complying with EU, US and Japanese standards. The company operates a modern, fully irrigated plant nursery and conducts trials of new and experimental crops. It maintains 25 head of cattle and an organic composting operation. UCIL processes sugar for the domestic market, and vanilla and cardamom for export to buyers around the world.

Surrounded by village communities of small farmers, the company also operates the Sezibwa Estate as a centre for innovation, development and training. With the support of donor partners, it provides extensive farmer development and training programmes. Concentrating on non-traditional and high-value crops, especially vanilla and spices, UCIL has successfully introduced new crops and trained farmers. Examples of projects include the replanting of old coffee plantations with new coffee hybrids that are more productive and more resistant to disease, and the introduction of cardamom as a new, high-value crop to be grown as a complement to vanilla.

UCIL has outsourced much of its supply chain, especially for vanilla and cardamom, and now works with some 3000 small farmers. At present, UCIL is also in the early stages of introducing black pepper in Uganda and training farmers to become first-level processors in the value chain.

**Economic challenges: vanilla and cardamom**
*Vanilla* is a complex product with considerable natural variations, and the vanilla market is equally complex. There is no auction system or transparent open market for vanilla, nor, generally, are there long-term supply contracts. Over the last decade or so, export prices for vanilla have ranged from US$15 to US$500 (€11 €370) per kilo, depending on its origin, quality and other factors. This price volatility places pressure on every level of the production chain, from small farmers to end users. For example, when prices fell drastically in 2004, small

farmers lost approximately 95% of their revenues from growing vanilla. At the other end of the chain, when prices peaked in 2002 2003, large industrial users switched away from using natural vanilla towards synthetic alternatives, a trend that has continued.

For processors/exporters such as UCIL, the main economic sustainability challenge in the vanilla market lies in accommodating such volatility while preserving secure and reliable supplies from small farmers on the one hand, and product quality, reliability and added value on the other. Meeting this challenge requires a careful reading of the market, prudent purchasing decisions, production flexibility and diversification, and building a good reputation in the market along with a strong and loyal customer base.

The economic challenges for UCIL's *cardamom* operations reflect those for vanilla. UCIL introduced cardamom as an alternative crop for vanilla farmers (who were losing revenues) as a way to diversify their operations and income streams. The cardamom programme has been running for about six years, although overall production volumes are not yet sufficient for viable commercial export. UCIL is faced with the dual task of continuing to expand the production base, while targeting higher-yielding niche markets, such as fragrance applications, or direct retail, for the small quantities of cardamom currently being produced.

**Farmer-level considerations**
UCIL's donor-assisted work with small farmers has aimed at creating an economically diversified and environmentally sustainable production base for

high-value, non-traditional crops among Uganda's smallholders. In our view, an economically diversified, secure income stream for farmers goes hand in hand with sustainable and environmentally responsible growing practices.

UCIL has therefore focused on a number of key objectives at the farmer level. One example is extension services and organic certification. The company devotes considerable resources to providing ongoing training and organizing farmers, and to promote organic farming and certification. UCIL visits farmers every month to appraise their fields and growing practices, and organizes regular workshops and training sessions, which are supplemented by radio talk shows where farmers can phone in their questions.

Outgrowers are selected on the basis of a number of criteria, including their proximity to UCIL and to other producers, their receptiveness to growing organic products, whether they are complying with UCIL's advice on husbandry, their potential to be demonstration or 'model' farmers (i.e. influence in the community) or to be lead farmers and create organized groups, their ability to produce a portfolio of crops, and their capacity to move towards downstream processing. Despite this careful selection process, the farmer-level initiatives face numerous hurdles in the Ugandan context. Chief among these is the issue of compliance – that is to say, the receptiveness and consistency of target farmers to accept and implement the husbandry protocols that UCIL seeks to encourage.

Farmers do not always wish to adopt measures that would lead to long-term, sustainable success, for a number of reasons, including economics, opportunism and outlook.

- *Economics*. Although some parts of rural Uganda are reasonably productive and prosperous, the general level of poverty in many areas is appalling. The typical small farmer survives on a subsistence basis, with any cash crops serving only to supplement the subsistence lifestyle. When one adds to this cultural (clan members have a moral claim on a successful farmer's cash income) and demographic factors ( many farmers have several wives, each of whom may have up to seven children), it is easy to see that immediate needs easily outweigh any prospect of long-term planning, prudent management and delayed gratification. Farmers therefore sell their produce on the side whenever the opportunity presents itself, and this 'side-selling' results in financial losses to UCIL in the short to medium term.

- *Opportunism*. Closely related to such economic factors is a deep-set pattern of opportunism among farmers. The option of establishing long-term sustainable agriculture – through new crop varieties, diversification, more productive and profitable husbandry methods, the careful management of resources, and a longer-term approach to rewards – is given short shrift. Such opportunism is a significant obstacle to UCIL's efforts to promote a sustainable portfolio of spice crops in Uganda. UCIL guarantees that farmers can sell their produce. When demand is strong, the company tends not to access all production, due to side-selling. Conversely, when prices are weak, UCIL must buy all production that the farmers offer. Naturally, this causes financial losses in the short to medium term, and is one of main reasons for accepting donor aid.

- *Outlook*. The outlook and attitude of Ugandan smallholders also affect the success of UCIL's farmer-level initiatives. One factor is that levels of literacy and education in rural Uganda are generally low. Another factor is an attitude we call 'donorism'. Decades of donor interventions have resulted in expectations among farmers that farm inputs, advice, training and markets will be given to them, and that their success or failure is only tenuously connected to their own efforts. Thus, it is rare indeed to find a small farmer who is willing to purchase inputs, sees his farm as a business, takes the initiative and responsibility for his efforts, and perceives clearly the link between those efforts and ultimate success.

**Export and market challenges**

The company's main challenges occur at both ends of the spectrum: at the production level, to ensure sustainable supplies, and at the export level, to ensure sustainable market penetration, viable sales levels and customer satisfaction. UCIL must meet these challenges by leveraging the available resources and generating a profit at the bottom line.

The challenges at the export/market level can be grouped into three categories: economics, market structure and regulation.

- *Economics issues*. These encompass challenges driven by market forces of supply and demand, or by the realities of cost-effectiveness. For example, UCIL cannot move up the value chain for vanilla, such as by producing vanilla extract. Industrial users of vanilla extracts usually require blends of different origins, and it would not be cost-effective for UCIL to import ingredients from other vanilla-producing regions. Such basic economic facts must be accepted as limitations on what is possible for UCIL in terms of its export potential.

- *Market structure*. The structures and routines of established markets tend to limit the opportunities for developing countries to participate profitably, although this can be counteracted if actors in the chain work together. For example, one of the biggest challenges in the vanilla market is reliance on spot sales rather than long-term contracts. This increases the risk both for

exporters, who are not assured of sales, prices or profits when they decide on how much to produce and hold, and for importers/users, who cannot be sure of secure and steady supplies or their quality unless they choose to hold their own excess inventories. There is no reason why steady, long-term contracts – perhaps with built-in provisions for funding sustainability and development goals – cannot be put in place for vanilla.

- *Regulation issues*. The regulations on food products to be imported into the European Union are extensive and complex. While one can certainly understand the need for regulations in the wake of food crises such as BSE (mad cow disease), for example, EU regulations nevertheless make it very difficult for small and medium exporters in developing countries to enter the market. Accessing the relevant regulations, keeping pace with changes, interpreting them, and applying them fully is usually not within the reach of medium-sized exporters in Africa.

**Building sustainable spice chains**

The private sector, a wide array of development partners and, indeed, ordinary consumers have begun to recognize that the free market can not on its own deliver development and sustainability results.

One of the most successful development and production models in the spice sector of which UCIL is aware is that adopted in the Indian state of Kerala. After independence in 1947, successive leftist state governments in Kerala were able to put in place a state-run system for spice production which, over time, has enabled Kerala to become a recognized leader in the sector. The hallmarks of this system include price stability and guaranteed markets at the farmer level, and state-run research and development system, the results of which disseminated via committed extension workers. Farmers now produce a wide array of compatible crops, helping to ensure multiple income streams and diversifying risk, and are encouraged to engage in processing their own produce. The government maintains overall price stability and quality standards, and the state marketing boards are engaged in aggressive export marketing. Quite apart from the success achieved in spice production, this government approach has also helped to foster general development. Kerala became, and remains, one of the most developed of the Indian states, with basic development indicators such as literacy levels, female participation in the workforce, income equality, etc., that are higher than anywhere else.

Sustainability and equitable production carry a real cost. The regulatory environment and market structures around sourcing from a country like Uganda are heavily weighted against sustainability and equity. For example, organic certification is built around EU requirements, and the only parties guaranteed a return are the certifiers and consultants.

As another example, importers tend to shop around for the best prices, but in the process drive down prices and, in the long run, sustainability. European consumers are probably unaware of these issues, and highlighting them might benefit any company that takes the initiative to do so. In addition, over the long term, steady supply contracts that address these issues may provide more secure and reliable sources of supply for the companies who enter into them.

In UCIL's view, economic viability and profitable trade must be the cornerstones of long-term sustainability. The profits generated in the chain must be shared more equitably among the different levels involved, from the smallholder producers to the ultimate consumers. Within such a framework of equitable sharing, resources must be devoted to education, development, prudent resource management, and, overall, long-term sustainability.

For more information, visit www.ucil-uganda.com

## 4.6 Discussion

Producing and sourcing spices for high-quality or niche markets is not easy. Food quality and safety requirements are strict and, together with consumer demand for environmental and social friendly products, spice chains must be transparent and sustainable. At the same time such markets offer opportunities, as the demand for spicy foods and for natural and organic products is rising.

This chapter has examined the sustainability issues that producers, traders and exporters are currently attempting to deal with, and how they have seized opportunities offered by the spice trade. From the case studies presented here, what lessons can we draw and apply when designing strategies to support sustainability in the spice sector?

**Perceptions of sustainability**

Discussions of sustainability always run into the problem of conceptualizing and defining 'sustainability'. What do we mean? The five case studies presented here illustrate clearly that that the various actors in spice chains have quite different perceptions of what sustainability actually means. Western buyers, for example, may think that a product is sustainable if the production process is organic. But, as the case of PBP reflects, this might mean that producers who do not use pesticides are more exposed to insect-borne diseases. What is perceived as sustainable by the consumer is not always experienced as such by the producer. Trajectories towards sustainable chains should always start by investigating what sustainability means for the different actors involved.

**Systems of standards**

Western markets have set high standards for food quality and for the social and environmental conditions of the producers. Some companies regard such standards as barriers and shift to less demanding markets; others, like Laxmi Enterprises, believe that being able to comply with strict standards offers opportunities. A complaint shared by many buyers is that the systems of standards are not sound, quality checks are not carried out consistently, and that regulations can quickly change. Existing systems do not offer constructive support for producers to enable them to move towards sustainability.

Systems of standards can have an impact on the structure and the governance of the various spice chains. In order to be able to sell to markets with the most demanding standards systems it is essential that the different activities in a chain are in alignment. Actors along each chain have to be willing to share information, and functioning systems are needed to communicate new requirements and changes in regulations. This requires mutual trust and firm relationships, ingredients that may be novel to traders who are used to dealing on spot markets.

### Building relationships

To improve communications and commitment, closer relationships between the various actors involved in the spice sector are needed. Most relationships today are arms-length, with buyers and producers engaging in spot-market transactions, without contracts or guarantees to purchase in the future. Such arms-length chains are non-transparent and do not provide the firm foundation that is essential to ensure high-quality and sustainable supplies. Sustainable business require long term relations if they are to continue to produce safe, high-quality and sustainable spices.

For many spice farmers, their main concern is daily survival, which explains their economic behaviour. They often prioritize short-term gain, and thus opt for a higher price now, rather than a steady income over the long term, by selling their produce to a buyer who might not offer the highest price but may guarantee to purchase in the future. As Samash Nathu of Uganda Crop Industries Limited (UCIL) explains, such short-term thinking is the result of many factors, such as family claims on farmers' incomes or the high incidence of polygamy.

It seems to be difficult to bind producers and buyers. Producers have a short-term perspective, and for buyers it can be cheaper to buy on the spot. Among the five case studies, Elephant Pepper is the only one where the buyer does not deal with producers who 'side-sell' to local traders. Producers do not have the opportunity to sell to other buyers; Elephant Pepper requires only fresh peppers, so it is impossible for producers to stockpile their produce and look for other markets.

In all of the cases, prices are important for producers. However, receiving a good price each year is not the same as having a good relationship. Michael Gravina of Elephant Pepper believes that to establish long-term relationships with producers, buyers have to be credible and consistent – a farmer will only start to trust a buyer if he keeps his word year after year. Buyers should recognize the short-term thinking of many producers and respond to it, such as by paying cash on delivery. Patrick Bartelemy reports that Cassia Co-op has introduced a scheme that allows the producers to benefit from the value added to the cassia by company. They are paid the local price directly, and also receive a bonus three months later if the international price is higher.

Rather than teaching farmers how to meet the buyers' requirements, it is more important that buyers show that they are willing to engage with them. Such engagement can be achieved through field visits, but also by supporting the producers and their families, such as through community development projects.

Some companies, including UCIL and Cassia Co-op, believe that 'shorter' spice chains, with more direct contact between buyers and producers, would help to improve farmers' incomes. However, in most cases, buyers at the end of the chain have no contact with producers, but acquire their produce from traders and other middlemen. As Bharat Maskai of Laxmi Enterprises explains, direct contact with producers is often difficult, either because they live in remote areas or because the transport infrastructure is poor, so supporting high-quality, safe and sustainable spices requires good relationships with these traders. Traders need to be involved in quality control as they are often engaged in activities like sorting, cleaning, segregating or packing.

## Investment in relationships

All buyers confirm that long-term relationships with producers can be very costly. Being engaged with producers means, as Michael Gravina of Elephant Pepper puts it, visiting the field and staying in contact, even though this may involve not only transport costs but also a lot of time.

All of the enterprises presented here are involved in projects to strengthen the capacity of the producers from whom they buy. This support may be direct, such as by providing training, demonstration plots, etc., or indirect, through investments in schools and community projects. All the cases emphasize the importance of such investments, but also that all of them need external funds to do so. Laxmi Enterprises receives assistance from local institutions, both public and otherwise, UCIL from international NGOs, while Elephant Pepper is supported by a sustainable investment fund. Perhaps 'trade *and* aid' would be better slogan than the more familiar 'trade not aid'?

For the stakeholders involved in organic food chains, long-term relations are even more important. For organic produce it is necessary that a complete production area is devoted to organic. Both Laxmi Enterprises and UCIL maintain that shifting to organic spices requires investment. Obtaining organic or fair trade certification is expensive, as is training illiterate producers.

Moving towards sustainability should be a collective effort involving all actors in a chain. The following chapter examines the opportunities and the challenges faced by the food and retail sector.

# 5 The food and retail sector

Most spices are processed and incorporated into other products, such as bakery products, soups, sauces and snacks. Although spices are ingredients in a wide range of products, for most food and retail companies they are not yet critical products in terms of sustainability issues, unlike fish, cocoa, palm oil and coffee, for example. However, if a product is claimed to have been sustainably produced, then all ingredients must be traceable and originate from a sustainable source. In such cases, sustainability standards also apply to the spices they contain.

Sustainability requirements are being applied to a growing number of products, and a wide range of certification systems have been introduced to demonstrate their level of sustainability. While many of these are general systems such as fair trade or organic, some companies have developed their own private certification systems.

This chapter examines the growing demand from consumers for sustainable products and, in turn, the increasingly rigorous sustainability requirements of the food industry and retailers. It addresses how these trends offer both opportunities and challenges for the spice industry.

Section 5.1 presents the vision of Royal Ahold, one of world's largest retailers. Section 5.2 presents an assessment of the opportunities and challenges for the spice sector, based on the discussions during the Amsterdam conference.

## 5.1 Royal Ahold

*Karin Bogaers, Coordinator Social Compliance, Royal Ahold Group, Amsterdam, the Netherlands*

Royal Ahold is an international group of quality supermarkets based in the United States and Europe that provides an easy, convenient and appealing shopping experience through its customer focus. Ahold is committed to offering its customers value, quality and healthy choices, while building value for its shareholders.

Ahold's mission is to make it easy for customers to choose the best. It does this through strong local brands and by putting the customer at the heart of every decision. The company strives to provide the best products in a relevant range, the best quality, the best prices, and the best choices for a healthy lifestyle. To fulfil this mission, the company sources corporate brand products from thousands of suppliers in Africa, the Americas, Asia and Europe through interconnected supply chains.

## Corporate responsibility

Corporate responsibility is fully integrated into Ahold's everyday business. In all operations, the company is committed to helping its customers understand how the choices they make impact their health, the environment and communities, so that they can make informed choices. In addition, the company actively engages with employees, suppliers, shareholders, NGOs and other stakeholders for a sustainable future, and has adopted the 'triple P' model to ensure a good balance between the interests of people, the planet and profit.

The company's corporate responsibility strategy focuses on the four areas where it can make the biggest positive impact: healthy living, sustainable trade, climate action and community engagement. These four areas have been at the core of the corporate responsibility strategy since 2007 because they are relevant to the business and supply chain, and are important to all stakeholders.

## Environmental issues

Ahold supports the global efforts to tackle climate change and is determined to operate its businesses in an efficient way that limits environmental impact. The priorities are to minimize the environmental footprint by reducing energy usage, greenhouse gas emissions, water consumption and waste, and by supporting the conservation of biodiversity. As a global retailer, Ahold sources agricultural products worldwide, so that changing weather patterns and climate conditions could impact future farming operations, crops and food supplies. In terms of climate action, the company therefore focuses on three primary areas: its own operations, the products it sells, and communication with customers about their footprint.

The production of all the goods sold in stores has an impact on the environment. As a preliminary step towards increased environmental compliance, suppliers are encouraged to be open and transparent about their production processes and to explore ways to reduce their environmental impact.

To help tackle the issue of environmental impact in the supply chain, the company recently joined the Sustainability Consortium, an international organization consisting of retailers, consumer goods producers, universities, research institutes, government bodies and NGOs, in life cycle analyses. The consortium is working to improve the science of analysis and developing strategies and tools to assess the sustainability impact of products. Ahold has joined this and other similar groups to collaborate on finding comprehensive solutions for environmental compliance and to achieve results more quickly.

One of the major challenges is to find solutions that can be applied to different types of suppliers and supply chains, and which cover not just the full range of environmental issues, but also the entire product range.

## Social issues

Ahold is committed to ensuring responsible behaviour at each step in the supply chain. This includes the environmental footprint, but also how workers are treated. Because the assortment includes products sourced from all over the world, the company insists that all suppliers respect the rights of their employees and comply with labour laws, regardless of their location. It is also committed to helping suppliers grow their businesses in a responsible and sustainable way. One example is the Albert Heijn Foundation, where Ahold partners with NGOs and suppliers to implement projects on healthcare and education.

The Ahold Standards of Engagement outline requirements for social compliance, including compliance with national labour legislation and the Core Conventions of the International Labour Organization (ILO). The standards also require all suppliers in high-risk countries to commit to the Business Social Compliance Initiative (BSCI) or equivalent programmes. The company works with the BSCI to raise awareness among suppliers on compliance issues, and the BSCI organizes roundtables to engage local stakeholders on improving working conditions. Most of its efforts in sustainable trade focus on the products sold in the stores, rather than the ingredients that make up these products. For certain critical commodities – including coffee, tea, cocoa, some types of seafood, palm oil and soy – the company is engaged in initiatives targeted at the parts of those supply chains where there are recognized social or environmental problems, and works with UTZ Certified, the Round Table on Responsible Soy (RTRS), the Roundtable on Sustainable Palm Oil (RSPO) and the World Wildlife Fund (WWF)

## Sustainability ambitions

The goal of Ahold's corporate responsibility efforts is to make a difference and to benefit all stakeholders. In the areas of sustainable trade and climate action, our ambitions are:

- to improve the ecological and environmental footprint by making operations more efficient, by reducing CO2 emissions by 20% per square metre of sales area by 2015 against the baseline published in the 2008 corporate responsibility report;
- to encourage suppliers and customers to act in an environmentally responsible manner;
- to develop environmentally responsible strategies that reduce waste and conserve water;

- to achieve Global Food Safety Initiative (GFSI) certification for all corporate brand food suppliers; and
- to conduct BSCI or equivalent social audits on all suppliers of corporate brand food and non-food products in high-risk countries.

**Sustainability challenges**

As with any complicated issue, Ahold also faces several challenges when it comes to sustainability. Tackling environmental impact and climate change in the supply chain requires a retail company to address a wide range of issues, from biodiversity to greenhouse gas emissions. At the same time, certain environmental issues are more relevant for a specific product or chain than for others, and vice versa. The challenge is to establish and reduce its overall environmental impact in a complicated supply chain, to cooperate with suppliers and, most important, to explain those efforts in understandable terms to stakeholders, including customers.

In terms of social compliance and critical commodities, one of the major challenges is to increase the level of transparency in sometimes complicated supply chains, especially the commodity chains. Important steps have been taken in recent years, but the product range continues to develop and new insights require constant examination of supply chains.

Efforts for people and the planet must be balanced with profitability to ensure the economic viability of supply chains. That is why Ahold partners with suppliers in Africa and undertakes efforts to connect suppliers with organizations that provide training to develop their capacity. The company is also engaged in international efforts to work with other retailers and harmonize sustainability requirements wherever possible in order to avoid duplication of auditing and certification, and thus make life easier for suppliers.

Although many large retailers and manufacturers now see (parts of) the sustainability agenda as a non-competitive issue – and realize the potential for cooperation and sustainable impact – it takes time to analyze all relevant issues and develop appropriate solutions. This is particularly true because each retailer and manufacturer has its own set of relevant stakeholders (including consumer markets) that might have different interests or priorities.

For more information, visit www.ahold.com

## 5.2 Discussion

In response to the increasing demand for natural and healthy foods in general, and for spicy and exotic foods in particular, food and retail companies are introducing sustainable sourcing policies. Ahold, for example, requires its suppliers to obtain GlobalGAP[4] and BSCI certification. A company's strategy to achieve sustainability in its supply chains may be driven by market opportunities, or is enforced by regulation, or is pushed by consumers and NGOs. What is certain is that the food industry has an important role to play in supporting sustainability as it has power to set standards.

This chapter has examined the position of the food and retail sector. What have we learned, and what is important to know when developing a trajectory towards sustainable spice chains?

**Spices not strategic**

For most food and retail companies, spices are not on their shortlist of strategic inputs, unlike palm oil or cocoa. Spices usually are just one of many ingredients in a food product, so they are not seen as posing a risk, in terms of loss of market share, due to food quality/safety, environmental or social injustices at the production level. As a result, investments by food and retail companies in developing a sustainable spice industry are currently minimal.

To motivate the food industry to support sustainable spice chains, increased awareness of the spice sector is necessary. There are many reasons to enhance sustainability; spices are used to provide flavour, colour or fragrance in a wide

---

4   GlobalGAP is a private sector body that sets voluntary standards for the certification of good agricultural practice around the world (see box on page XX).

range of products. The opportunity costs of a decline in the spice industry are huge; the loss of the spice sector would be irreversible, as many spice plants take years to mature and bear fruit, and could result in a loss of biodiversity.

But who should take the lead in supporting the industry to move towards sustainability – the major retailers, governments or consumers?

**The consumer as driver**
First, the power of the consumer with regard to sustainability is debatable. Indeed, it is the consumer who, by demanding and showing willingness to pay more for organic or fairtrade products, is able to push the retail sector to take action. Through consumer groups and forums, consumers can force industry to provide sustainable products. Consumer power has its limits, however; consumers can support the food industry to move towards sustainability, but they can not assess or control it.

Moreover most consumers are unaware of the spices they consume. Spices are insufficiently visible in the market and so there is little social pressure on the industry for change. Enhanced consumer awareness would provide opportunities for industry brand and sell 'sustainable' spices.

**Collaboration**
To enhance sustainability collaboration is necessary, both *horizontally*, between retailers, to harmonize standards and certification schemes, and *vertically*, to ensure that all actors within a chain share responsibility for sustainability.

According to Karin Bogaers of Ahold there is little cooperation between retailers with regard to sustainability. The requirements of countries and retailers can vary enormously. There have been moves to improve cooperation, however, and the Global Food Safety Initiative (GFSI) has been launched to function as an umbrella for all certification schemes. By functioning as a benchmark for existing certification schemes, this initiative aims to reduce the number of schemes and thus improve transparency for suppliers and consumers. There could also be more cooperation among retailers with regard to social auditing. Because producers sell to different companies, they are subject to several but very similar auditing processes, all of which are costly and time consuming.

There is also a need for greater vertical cooperation. At present, retailers do not have the power to enforce sustainability in the rest of the supply chain. They can influence the agenda of first-tier suppliers, but the larger the distance between actors in the chain, the more difficult it becomes to control and share responsibility.

**Costs**
Even though it is still widely assumed that sustainability costs money, more and more retailers are recognizing that investing in sustainability can lead to increased efficiency in the long term. Using energy and water more efficiently, and minimizing $CO_2$ emissions can help to reduce costs. While it may be difficult to measure the impacts of greater efficiency on social sustainability, it may be assumed that healthier, satisfied producers will also produce more, better-quality produce.

**Priorities**
From saving energy in logistics to paying fair prices to producers, it appears that moving towards sustainability will be a long process. Where do we start? As the impact of actions may not always be apparent or even definable, many actors may find it difficult to prioritize the different activities that support sustainability. Assistance in designing action plans is needed.
In the Netherlands, a few individual companies have launched their own sustainability initiatives, , but most have originated in some form of public-private cooperation. Although the private sector must be at least willing to consider engaging in moves towards sustainability, other parties are needed to pave the way, to co-invest in designing initiatives or to provide moral support. The next chapter examines the role of the public sector.

# 6 Regulations and certification

The different actors in spice supply chains deal with different issues, ranging from sustainable production to dealing with demanding consumers. Each chain as a whole is influenced by the policy context at local, national and global levels. Food safety regulations for importing spices into the European Union and the United States are known to be tough. On top of that, EU and US markets are requiring higher social and environmental standards in the exporting countries, so that greater transparency in supply chains is essential.

Can certification systems be useful tools for managing food safety regulations and sustainability standards? How do governments perceive sustainability in supply chains and what future policy can be expected? To answer these questions, this chapter includes the presentations of John Fagan of the Earth Open Source Institute, and Alex Bruijnis and Tjeerd de Vries of the Netherlands Ministry of Economic Affairs, Agriculture and Innovation. They share their visions of sustainable chains and consider the opportunities and challenges that regulation and certification can bring.

## 6.1 Certification for a sustainable world

John Fagan, Earth Open Source Institute, USA

### Certification and standards

Certification is a useful tool for third-party verification of the practices, procedures and policies involved in delivering a wide variety of products and services. Certification itself is a relatively straightforward process. It involves third party inspection of documents and operations, systematically delivering reports regarding these items, and determining whether an operation or product complies with the certification standard. There are many certification schemes, based on standards and systems established by governments, industry and private companies.

*Government standards.* The standards created by government are of two kinds. *Mandatory legal standards* require retailers that wish to sell a certain kind of product to demonstrate, on the basis of an independent assessment, that it meets a certain set of requirements. *Voluntary legal standards* mean that

*First pathway*

An issue arises in society, resulting in a broad discussion and the elaboration of many lines of action related to that issue. One of those lines of action is that one or more organizations, either profit-making or non-profit, will create standards relevant to the issue and structure a certification programme based on those standards. Depending on many different factors, these private standards may either become widely adopted by industry, or industry may opt to develop their own standard or standards. In either case, over a period of time the private/industry standard may become widely adopted within the industry and, if it is a consumer-based system, it may become widely relied on by private citizens. The result is that practices relevant to the issue will have evolved as a result of the certification standard and its wide adoption within the industry and acceptance by citizens. Typically, when change has been established, government may decide to step in and codify the standard in a law or regulation. This is exactly what happened with organic agriculture.

There is discussion regarding whether a similar route should be taken with social and environmental certification. The argument is that there are now so many programmes that consumers are confused, and that a single, government-mandated voluntary certification programme would be more desirable. The European Union already has the EU Ecolabel scheme, a sustainability certification programme that applies to non-food, non-agricultural products. Some believe that it could be expanded to cover food.

*Second pathway*
An issue arises in society, resulting in a broad discussion and the elaboration of many lines of action related to that issue. The government announces that it has decided to create laws or regulations governing the domain of the issue of interest. One of two things may happen:
- Either industry steps in and says, 'we will self-regulate in this area'. Then industry creates a standard and certification programme designed to ensure that all industry members operating in that sector are complying with the standard and are thus addressing the issue of interest in an acceptable manner. Often, even when industry initially sets up a standard, the government will step in at a later time and implement regulations.
- Alternatively, the government establishes regulations governing the issue and sets up a system for enforcing them. One system of enforcement that the government can mandate is a certification programme that verifies that industry is complying with the regulations.

## The transformative role of private standards
Typically, industry and government standards address the basics. They drive change, but only incrementally. Private standards often play a key transformative role.

Often seen is that a segment of society is strongly supportive of significant change. Private standards developers often plug into this demand and develop standards that assist that segment of the public in locating products that meet their more demanding values and requirements. To the extent that the public adopts such standards and uses them to guide their purchasing habits, then private standards can drive significant changes in industry practices and ultimately in law.

A good example of this has been certification related to foods that are free of genetically modified organisms (GMOs). The European Union has issued a directive requiring that food products containing GMOs must be labelled as such. This resulted in the *de facto* exclusion of GMOs from foods sold in most European countries. The EU directive did not, however, require labelling of products derived from animals raised on feed containing GMOs. The CertID Non-GMO Certification Programme was created to support brand owners and retailers in offering non-GMO foods. But, recognizing that consumers were also interested in non-GMO animal products, the programme was extended to certify animal feed. This enabled retailers in particular to offer non-GMO fed animal products. As the demand for these products grew, the governments of Germany and Austria responded with certification programmes providing a seal that applies to both non-GMO foods and to products from animals fed non-GMO feed. Similar programmes are in now progress in France, Ireland, and Italy. These activities have in turn led to discussions within the European Parliament calling for mandatory labelling of all products from animals raised on feed containing GMOs.

## Private standards and sustainability

In recent years, society-wide discussions of sustainability have been triggered by the recognition that humanity is confronting several very serious challenges. These challenges include climate disruption, energy depletion, poverty, hunger, population displacement, ecosystem degradation, water depletion, health threats and the destabilization of the global economy. These discussions have prompted both government and industry to introduce several sustainability certification initiatives. However, these initiatives are currently limited to motivating modest, incremental changes. A cursory analysis leads to the conclusion that the scope of change that these programmes motivate is simply too limited to contribute significantly to solving the major problems we now face.

A rapidly growing segment of the population clearly recognizes that transformative change is required to address these problems effectively. Many people are looking to channel their resources to support companies that have adopted transformative, not incremental, approaches to sustainability, and to use their purchasing power to support services and products that are the result of such approaches.

This has created a new niche in the certification sector for programmes that provide incentives for transformative corporate behaviour that, if adopted by a sufficiently large proportion of industrial concerns, would contribute significantly to addressing the problems we face.

### EOS World Sustaining Certification Programme

The Earth Open Source standard is designed to address the challenges and opportunities described above. It is intended to motivate transformative corporate change leading to solutions to the sustainability challenges of today, and to enable those companies courageous enough to take on this leadership role to market their services and products effectively.

The EOS standard is built on the foundation of the more conventional ProTerra certification that Cert ID has offered since 2006, but it includes some novel elements. Companies will be admitted to the programme only if they comply with a few basic requirements that assure that their operations are at least free from egregious practices that would create significant public acceptance problems, either for the client or for EOS. They must also complete a sustainability assessment and planning process that:

- considers the company, its activities and its products from the perspective of 12 basic areas, referred to as the '12 vital signs' of sustainability;
- identifies specific long-term transformational targets for each of the 12 vital signs, the achievement of which would not only bring the company to a significant level of sustainability, but would also contribute to the sustainability of communities, nations and the world; and
- designs a plan of action and accomplishment, including specific intermediate targets for each of the 12 vital signs, and a plan and timeline for achieving them (see box).

For each company, ongoing participation in the programme is based on consistent progress towards fulfilling the plan of action and achieving the 12 transformational targets in a timely manner. It also requires regular third-party assessments to gauge progress towards each of these targets, and regular, transparent reporting. All participating companies must also be involved in a collective programme promoting their sustainability achievements, based on their accomplishments.

## Transformational targets

Examples of transformational targets for each of the 12 vital signs of sustainability include the following (actual targets would be set by each company):

Restoring a regenerative carbon cycle – climate change
- Corporate operations carbon neutral by 2015.
- Supply chain carbon neutral by 2030.

Regenerative energy systems
- 100% renewable energy for own operations by 2020.
- 100% renewable energy for supply chain by 2030.

Regenerative farming – soil, fertility and pest management
- No petroleum-based pesticides or fertilizers by 2020.
- On-farm and locally produced pest and fertility management by 2030.

Regeneration and protection of biodiversity
- Support suppliers in restoring and preserving damaged high value conservation areas and creating/restoring wildlife reserves by 2020.

Regenerative food systems – local food self-sufficiency
- Transition to local suppliers for all inputs possible by 2020.
- Support distant suppliers who are no longer used in transitioning to local regenerative food production in their region. Local self-sufficiency by 2030.

Regenerative water systems
- Reduce water use in manufacturing by 30% by 2015.
- Reduce water efficiency of all agricultural suppliers by 50% by 2030.

Regenerative materials management – recycling, waste management, packaging, conservation and life-cycle analysis
- Reduce packaging in consumer products by 50% by 2015.
- Own facilities – transition to entirely recyclable packaging by 2020.
- Suppliers – transition to entirely recyclable packaging by 2025.

Regenerative economy – economic development
- Co-create food-oriented local economic development projects by 2015.
- Co-create social ventures leading to local food self-sufficiency by 2030.

Regenerative communities – community development
- Co-create democratically run community centres, which will host education, cultural events and social ventures by 2020.

**Regenerative personnel management – optimized worker welfare**
- Continuing education programmes for all corporate employees by 2012.
- Continuing education programmes for all supplier's employees by 2015.
- Living wage for all workers in the full supply chain by 2020.

**Regenerative marketing – transparent communication with consumers**
- Promote products based on their real value to the buyer and based on the company's contributions to global sustainability by 2015.
- Design promotional activities to stimulate learning about sustainability and the need to achieve a globally regenerative food system by 2015.
- Fully transparent and complete labelling for all products by 2018.

**Regenerative products – quality, safety, nutrition and sustainability**
- Design/reformulate products and production systems to support the achievement of all the objectives outlined above.
- Design/reformulate products and production systems for maximum nutrition and wholesomeness by 2015.

The basic premises of the EOS programme are as follows:

*First mover advantage*. Companies that establish themselves as first movers in the area of transformational sustainability will reap significant benefits in the form of savings and economic opportunities opened up by undertaking transformational change, and in the form of recognition in the marketplace – consumer, employee and shareholder loyalty and enthusiasm for the company, its products and services.

*Corporate leadership for a sustainable future*. The scale of change required to establish a sustainable future for our planet is so large that it is beyond the scope of the resources available to either governments or the public. The only sector of society with sufficient resources to address these challenges is the corporate sector. Since transformational change led by the corporate sector is essential for ensuring the sustainability of our world for this and future generations, it is incumbent upon companies who have understood this fact to take a leadership role that will motivate and empower other companies to undertake transformational change themselves.

### The spice industry: an opportunity for leadership
The spice industry faces a number of truly unique sustainability challenges and opportunities. The complexity of sourcing, the length of supply chains, the diversity of participants in each chain, both culturally and economically, and the penetration of the spice chain into virtually every branch of the food system, all contribute to these challenges.

The food system is intimately connected to each of the major challenges of our time: climate, energy, poverty, hunger, ecosystem degradation, water, health and the economy. Transforming the food system can contribute substantially to addressing each of these challenges. For example, making the food system carbon neutral would contribute to addressing the problem of climate disruption by reducing global greenhouse gas emissions by 30 to 40%. Because of the far-reaching nature of many spice chains, improvements in their sustainability would have far-reaching impacts on the sustainability of the global food system.

Thus, if the leaders in the spice sector were to take a leadership role in modelling transformational sustainability programmes, their leadership could have a disproportionately large impact on the food system as a whole, leading to profound and world-changing transformation in the direction of real sustainability.

At the same time, leaders in the spice industry have already begun to implement the key enabling elements for undertaking a programme of transformational sustainability, such as achieving full traceability of their supply chains. Thus, they are more likely to be able to make significant progress in achieving transformational sustainability targets than would be the case in many other segments of the food system.

This is an opportunity for the spice industry to create new models for success in business, and new modes of operation that will sustain our world for future generations.

For more information, visit www.earthopensource.org

## 6.2 The role of governments

*Alex Bruijnis and Tjeerd de Vries, Ministry of Economic Affairs, Agriculture and Innovation (EL&I),[5] the Netherlands*

The UN predicts that in 2050 there will be more than nine billion people on this planet, three billion more than today. All will require food and access to drinking water and will have material needs. The current demand for food, water and goods is already putting a lot of strain on the world's production systems. The result is overexploitation of available natural resources, such as uncontrolled logging of tropical forests. An alternative, sustainable path is clearly needed.

The European Union is the world's biggest importer of spices, and within it the Netherlands is one of the main importers and re-exporters. This means that the

---

5   Formerly the Netherlands Ministry of Agriculture, Nature and Food Quality (LNV).

Netherlands has a special responsibility to address the challenge of ensuring the sustainability of spice production and supply chains.

Until now, the sustainability of herbs and spices has received little public attention, nor has it been a focus for the Netherlands Ministry of Economic Affairs, Agriculture and Innovation (EL&I). While a wealth of information is available on spices, very little of it concerns sustainability. Are there no sustainability issues in the spice trade? Or are spice producers and traders seen as minor players compared with those involved in supplying timber, soy and palm oil? Behind the scenes, however, the sector itself has the ambition to become sustainable. The Dutch Spice Trade Association (NVS) recently approached the Dutch Sustainable Trade Initiative (IDH) to develop a sustainability programme for the sector. This section looks at the Ministry's position on sustainability and what role the Netherlands government can play in the programme.

**Sustainability issues are here to stay**
As consumers become more aware of sustainability issues, and become convinced that changes need to be made, it is expected that they will increasingly demand products that are produced with care for people and the environment. As yet, such products are supplied to niche markets, but the social and environmental values behind them are expected to continue to spread and become mainstream. And societal demands will continue to evolve in response to new challenges that emerge.

What are the consequences of this for entrepreneurs? Producers and traders need to operate in accordance with the values of the society of which they are part. Whether they are at the forefront or at the rear of the movement towards sustainability, they will have to take action and adapt. Sustainability will become an economic necessity.

**A look at sustainability in a legal and social framework**
Sustainability is a process where stakeholders develop, step by step, profitable production and consumption chains based on a foundation of care for people and the environment. This process has no fixed outcome, or a clear 'end goal'. Sustainability puts demands on the product and its lifecycle. It concerns the complete value chain: all the steps in the chain from the soil to the end product. From a government's point of view, these demands for sustainability can be seen at two levels:
- *Level 1*. Basic requirements that are enforced by laws and regulations. The role of government is to set the basic constraints for production (and trade) by laws and regulations. In the herbs and spices sector, a large part of the regulations are aimed at food safety. Entrepreneurs in the sector will experience the demands from such regulations in their day-to-day work.

Other examples include general regulations regarding the use of pesticides or laws prohibiting child labour.
- *Level 2.* Demands on products that are not enforced by law, but are demanded by society. Producers have to meet those demands that are above the minimum legal requirements either on a voluntary basis, or they will be forced to do so by the market, requiring compliance with standards ranging from good agricultural practice, such as GlobalGAP, to sustainability standards.

The role of government is to encourage and facilitate parties who wish to move towards more sustainable products, using a variety of instruments such as providing funding for research or participating in public–private initiatives. With greater knowledge and transparency, both producers and consumers are better informed and are able to make sound decisions. The Ministry, for example, provides funding for the Dutch Sustainable Trade Initiative and international roundtables (see below). Other instruments include measures to encourage new investments with subsidies or to share risks, as explained in the following.

**CSR and sustainable trade**
Since the first level of sustainability is obligatory (laws and regulations), this section focuses on the second level of market-driven change. With respect to the herbs and spice industry, two areas can be distinguished: corporate social responsibility (CSR) and sustainable trade.

CSR is about companies taking responsibility for the negative side effects caused by their operations. CSR issues include: does the company operate in a socially responsible way, pay fair wages, efficiently manage its use of energy ($CO_2$ emissions) and water, etc. Since each company determines which social values it addresses, and how, there is no absolute normative standard for 'good CSR'. However, companies do have the responsibility to be transparent about their CSR initiatives and their results, and describe progress in their annual reports.

CSR can spread from the company through its value chain or network to other companies or producers. For companies in spices and herbs, this will involve trade relations. Because of this, sustainable trade is often part of a company's CSR strategy, focusing on the production of and trade in raw materials. Here, the main issues include, for example, ensuring that the company provides acceptable social and working conditions for its workers, or its management of natural resources is adequate. Again, there is no general normative standard to distinguish between sustainable and unsustainable trade. But at the commodity level, generally accepted (normative) standards can emerge to determine what is considered to be sustainable. This can be achieved via international cooperation, such as the roundtables on sustainable palm oil and soy production, which are partly funded by the Ministry (see below).

The Ministry encourages CSR and sustainable trade within the concept of 'triple P': people, planet and profit. The last P is by no means negligible: the transition to more sustainable production and trade has to benefit the companies involved. The benefits may be immediate, such as through increased efficiency savings. Companies can also benefit in the long run because they have access to high-quality inputs, and because they are able to stay in the market by responding to the changing demands of consumers and retailers for sustainable products.

**International cooperation**
The globalization of business means that international cooperation is necessary. The Netherlands government encourages international partnerships with stakeholders such as producers, traders and NGOs (consumer groups) to increase the sustainability of trade and production. The Roundtable on Sustainable Palm Oil (RSPO) and the Round Table on Responsible Soy (RTRS), for example, involve step-by-step processes that follow a general pattern:
1 key issues are identified,
2 criteria are set for these issues,
3 levels or norms are set for the criteria,
4 compliance is ensured by certification and labelling.

The process usually starts with a limited number of parties, with more parties joining the process at each step. Some parties also drop out, of course. At every step, people with sometimes opposing points of view need to reach an agreement, which often makes the negotiations very complex and time consuming. The result might be an agreement on how to produce, transport and use guaranteed sustainable soy, for example. Once an agreement has been reached, the producers and industry have to incorporate it into their day-to-day business.

**Business and consumers are responsible for further sustainability**
The Ministry believes that producers and consumers themselves are responsible for improvements in sustainability. Producers, traders and retailers are key to changing the system. The government plays a facilitating role in this process and functions as a partner. This is not a fixed or predetermined role. Sustainability problems vary and depend on many factors such as the intended use of a commodity, where it is produced, or how it is transported. Likewise, the role of the government can differ to fit the circumstances. Several subsidy schemes are available for sustainable development, some of which might apply to the sustainability initiative in the herbs and spices sector. When specific points of action for sustainability emerge in the sector, the Ministry can help to address these topics. The production chains for herbs and spices are very complex, however, and there are no easy solutions. But step by step, sustainable change can be achieved.

**Too much of a good thing? Certification and labels**
There remains the issue of how to guarantee that a product is really all that it claims to be. In the past, inflated claims concerning sustainability have emerged (referred to as 'green washing'). Consumers have naturally become suspicious of these claims and of the overabundance of 'green' labels. Making credible claims is vital to maintaining consumer confidence. A popular solution is the certification of production and trade. Through labelling, the producer can communicate that its production methods satisfy a number of requirements,

which are checked by audits. On the downside, the market for labels has grown tremendously in recent years, resulting in such a wide array of labels that only well-informed consumers can differentiate between them. Another consequence is that producers are asked to comply with a bewildering array of standards for different labels. As a result, there have been requests for government intervention.

The Netherlands government is in favour of transparency and clear communication to consumers and would like to see a reduction in the number of labels. This is a point of attention and action for the government, but labels are currently privately owned and World Trade Organization (WTO) regulations leave no room for government intervention. The government can, however, decide on which labels it chooses for its own procurement, and can also choose certified (and hence labelled) materials wherever subsidies are involved. This is the case for the obligatory addition of biofuel to gasoline and diesel fuels in Europe. This is considered a subsidy, and therefore only certified biofuels are used. In all cases, transparency is key: consumer labels need to make clear which goals they support and what the results are.

Sustainability is one of the key issues for the future. Resolving it will be a step-by-step process in which producers, traders, consumers and governments all share responsibility.

For more information, visit www.minlnv.nl

## 6.3 Discussion

Standards and certification can help in improving the sustainability of the food supply system. This section sets out the opportunities and the challenges offered by standards, regulations and certification for spice production and trade, and examines the role of the government in sustainability in spice chains.

**The purpose of certification**
Sustainability and food quality and safety standards demand that the transactions in value chains are transparent. Certification is a mechanism that can help to restore transparency, accountability and confidence, benefiting both consumers who can 'trust' the products they buy, and sellers who can use the certification as a selling point.

Certification of spices could serve several objectives. For companies, it could provide proof of credibility, enabling them to demonstrate that they are complying with a specific set of requirements. It could also assist corporate or private buyers in making procurement decisions that are consistent with the specifications established by consensus within the industry, or established by private certification bodies and accepted by consumers as reflecting their personal values. It would assure the buyer that the spices meet agreed standards. And certification ensures the government that an industry is complying with relevant laws or regulations.

Despite the penetration of the spice chain into virtually every branch of the food system, there are as yet no specific industry standards for spices. As the sector can contribute broadly to sustainability, there is a need for leadership.

**Drivers of sustainability**
While government and industry certification initiatives provide only modest incentives for change, private standards can play a key transformative role. When a segment of society strongly supports significant change, private standards developers often plug into this demand and develop standards that assist that segment of the public in locating products that meet their more demanding values and requirements.

Standards and formal certification alone will not move the sector towards sustainability, however. Collaboration between leaders in the industry is much more important. Industry has the power to move the sector, set goals and communicate with the public.

The Netherlands Ministry of Economic Affairs, Agriculture and Innovation (EL&I) believes that producers and consumers themselves are responsible for improvements in spice sustainability, but that the government can help to facilitate this process and function as a partner. An example is the Dutch Sustainable Trade Initiative (IDH), a multi-stakeholder platform for improving the sustainability of international supply chains. Although IDH is funded mainly through the Netherlands Ministry of Foreign Affairs (Directorate General for Development Cooperation), the initiative is supported by all ministries.

**Setting standards**
If the aim is to set standards that support sustainable spice chains, where do we start? With the easiest, the biggest or the most pressing issue? The spice sector is enormously complex, involving many different products, actors and countries, so it is difficult to set general standards. John Fagan advises companies and other actors to analyze their own situation, 'start at home with you our own stakeholders, suppliers and employees', and bear in mind that moving towards sustainability is an ongoing process that cannot be captured by a certificate or label alone.

Achieving sustainability will be a long process that will necessarily involve transforming current structures. Sustainability is not a box that a company can tick and then issue a label. Sustainability reaches further than one company; it also includes the expectations of all the actors involved in many chains. It may be difficult to ensure integrity of purpose, and benchmarks can assist in that, but reaching such benchmarks alone is unlikely to transform existing structures.

Thus, if the leaders in the spice sector were to take a leadership role and provide models of transformational sustainability programmes, they could have a disproportionately large impact on the food system as a whole, leading to profound and world-changing transformation in the direction of real sustainability.

# 7 Managing relationships

As mentioned in the previous chapter, in order to guarantee supplies of spices to Western markets, solid relationships between the many actors involved in the various supply chains are essential. Increasingly, it is not just the actors themselves who are involved in developing sustainable supply chains, but also public stakeholders and service providers, such as financial institutions.

Cooperation between stakeholders with different objectives and priorities is not always easy. This chapter reflects on the challenges regarding collaboration between public and private sector actors.

## 7.1 Golden Food Products, Tanzania

*José Ruijter, programme officer, Cordaid, the Netherlands, and Cleopa Ayo, general manager, GFP, Arusha, Tanzania*

Golden Food Products (GFP) was established in 1998,[6] and is based in Arusha, Tanzania. The company processes, packages and sells 15 varieties of branded and attractively packaged products – including natural fruit jams, all-natural peanut butter and honey – directly to hotels, restaurants, tourism companies and supermarkets throughout the country. In 2006, GFP set up a business to produce organic spices for export. For this purpose, GFP has its own small farm and processing unit in the Usambara mountains, in Tanga region. The company has ten permanent staff, including four field officers and 40–80 casual staff who produce and process the spices. In addition, the company selected one lead farmer in each village to be trained to act as the contact for other farmers in the village.

### Mission
Under its latest five-year business plan (2011–2015), GFP intends to increase the number of importers of its organic spices in order to attain its vision of becoming a leading manufacturer and processor of high-quality organic spices for both local and export markets by using only natural ingredients.

---

6 GFP's current owners, Mr and Mrs Cleopa Ayo, both own 50% of the shares. Although GFP also produces jams, this case study focuses on its organic spices.

GFP produces various organic spices, such as black and white pepper, cardamom, lemon grass (used in teas), cloves, cinnamon and ginger. Quality management throughout the entire chain is crucial. In the export market, prices are determined by the quality of aspects such as wholeness and size of the spice seeds, colour and the content of essential oils. GFP's management is qualified to deal with organic quality markets: the company director has been trained in ISO 22000 (food safety management systems) and in HACCP. GFP has received organic certification from the Institute for Market Ecology (IMO) for four years in a row.

**Relationship with farmers**

GFP works with 625 farmers in six villages in Muheza district located at varying altitudes in the East Usambara mountains. The farmers, registered and contracted by GFP, produce organic spices, which they sell to the company. Because spices are perennial crops, a steady supply of produce should be possible as long as the company can control the interference of local traders who buy part of their produce. To address this 'side-selling', GFP has invested heavily in strengthening its relationship with the farmers by providing extension services in collaboration with third parties. With the support of Dutch donor agency Cordaid, GFP now has access to working capital, allowing the company to pay cash on delivery.

Over the last few years, also with funding from Cordaid, GFP has signed contracts with Kiwi Consultants, an international group, and Faida Market Link (Faida MaLi), a local NGO, to provide training for farmers. The training has focused on subjects such as farming as a business and record keeping, good cultivation practices, erosion control measures, weed and pest control, pruning, fertilizers and crop rotation. The Export Promotion of Organic Products from Africa (EPOPA) programme facilitated the introduction of organic certification. The farmers were certified through the Institute for Market Ecology (IMO), following an inspection by the Tanzania Organic Certification Association (TanCert). At Kwamhosi, a village at the centre of the growing area, GFP has established a collection point and processing facility, which will be further expanded. The famers have also received assistance from GFP to establish nurseries for growing seedlings so that they can expand the area under cultivation.

In 2008, with a grant of €67,800 from the United States African Development Foundation (USADF) the farmers established the East Usambara Organic Farmers' Association, called Juwakihuma. USADF helped the association prepare a business plan and financial manual, and provided training in organic farming techniques. In the medium term, Juwakihuma plans to take over some extension activities from GFP, thus reducing the company's extension costs and management burden. GFP is also involving the district authorities to try to address the growing problem of the theft of spices from the farms.

## Market

Tanzanian producers and exporters have established a solid reputation as sources of quality of organic spices at reasonable prices. The East Usambara mountains and the island of Zanzibar are among the best areas for organic spice production, making use of existing smallholder farming systems, traditional know-how, favourable soil and climate conditions and biodiversity. GFP offers first-grade products for export to the European market, with second-grade products destined for the regional (Kenyan) market, although its market share is currently very small, at less than 1%.

In 2007, GFP exported its first batch of spices to Europe through Doens Food Ingredients, a Dutch wholesale importer, but the company was not satisfied with the prices offered by Doens. In 2009, the company established collaboration with Organic Food Ingredients Sagl (OFI) in Switzerland and Diafoods, an Indian company, to export small quantities of spices to specialized niche markets in Europe.

OFI does not become legal owner of the goods, but performs a number of key tasks, including:
- acting as marketing agent for GFP (which does not yet have a marketing department[7]) against 30% of the net profit on organic products, plus expenses;
- identifying profitable niche markets in Europe and providing information on prices, market opportunities, delivery requirements and payment terms;
- screening clients' business credit ratings and handling payments; and
- facilitating the transfer of payments from clients to Diafood and onwards to GFP.

Diafood GmbH (Germany) and Diafood France (DFF) are the main importers/ distributors to the European market, and are together responsible for:
- organizing import permits, transportation and storage of goods at DFF's organic-certified warehouse facility in France (for which they charge a service fee of 3% of the cost of goods and an annual fee of €3000), while ownership of the goods remains with GFP; and
- paying levies, customs duties and insurance, as well as laboratory analysis and investigation/ legal defence (deducted from payments to GFP).

Currently, OFI's main customer is Sonnentor, a web-based organic specialty sales agent that sells direct to consumers. Sonnentor does not buy large quantities but calls for smaller quantities based on final delivery orders in order to minimize its financial and storage costs. This spot selling arrangement is time consuming and places a heavy burden on GFP's cash flow and cash management.

---

7   The company intends to set up a marketing department in the period 2011–2015.

In order to reduce the impact of middlemen who buy spices directly from the farmers at low prices, and thus reduce the volume of quality spices supplied to the company, GFP has contacted the Kenyan dealers who buy directly from the middlemen, and has tried to convince them that they can purchase higher-quality produce directly from GFP.

**Financial support to farmers and GFP**
As well as helping to establish the farmers' association, USADF provided GFP with a grant of €182,072 to invest in processing fruits and spices for the period 2010–2011. In return, GFP has promised to contribute €45,500 to community projects, although these are still to be defined.

Cordaid has also provided financial support through a combination of loans and grants for technical assistance. For example, it has helped GFP to meet with similar companies to exchange experiences, and with buyers at gatherings such as the BioFach, the World Organic Trade Fair held each year in Germany. In 2008, Cordaid provided a grant of €20,000 to fund training activities implemented by Faida MaLi and Kiwi Consultants for the spice farmers. Another grant is planned for 2010 2012 to increase the production capacity and farming practices of spice farmers and to continue building their trust and loyalty. Some technical assistance will be provided to help improve the training and inspection skills of GFP's extension staff, and the overall financial management capacity of the directors.

During the first season (2008–2009), Cordaid provided a loan of €100,000 as working capital, which was fully repaid. A new loan of €150,000 for working capital/ trade finance for the period 2010 2011 has just been approved, at an interest rate of 6% for a one-year term. Cordaid only provides financing to small and medium enterprises (SMEs) that work with outgrower/contract farmers who can not access other sources of funding, and who usually do not have sufficient collateral to obtain loans from banks (which require 125% collateral because of Central Bank regulations.)

**Ensuring the commitment of partners**

The partnership between the various actors in this value chain is based on a shared vision and the mutual benefits that all participants (chain actors as well as the support actors) gain from the collaboration. The interplay between key actors in the chain, from the farmers up to international buyers, supported by internal extension staff and external training providers and financial institutions, has led to the development of a spices value chain that is transparent and is gradually growing in terms of volume and profitability, and in terms of the social capital of farmers and institutional capacity of GFP.

The growing capacities of the farmers and their association, Juwakihuma, and of GFP itself, are creating the conditions for a more inclusive and transparent collaborative relationship. The training in farming as a business has helped the farmers to understand the costs and margins involved in the value chain, and has helped to build their trust in and loyalty to the company. Meanwhile, training in financial management is enabling GFP to carry out thorough analyses of costs and risks.

The farmers in the area have been growing spices for decades, but marketing was a significant problem for them. Many buyers came and went, as did middlemen from neighbouring countries, and farmers became wary of new buyers. By collaborating with GFP, farmers now have access to a secure market outlet for their spices, with fair contracts.

GFP also provided training and organizational support for the farmers. Through their linkages with GFP and the training offered through the cooperation between GFP extension staff and external training providers (including the Dutch Agro Eco–Louis Bolk Institute, Kiwi Consultants and Faida MaLi), the farmers are gradually building up their own association, which has allowed them to reduce transaction costs and to speak with one voice in their negotiations with the company and service providers. By applying their new skills in farming as a business and improved agricultural practices to their food crops, the farmers have improved their productivity and their incomes.

Through this collaboration, GFP should have access to steady supplies of quality spices. However, side-selling by farmers to local traders remains a key problem despite GFP's investments and energy. The company's strategy to address this has been to continue to invest in training for farmers – so that they become aware of the fair price that GFP is providing – and in enabling them to access finance. Gaining access to finance is key. Indeed, farmers are side-selling their produce to local traders not because they offer attractive prices, but because they need cash immediately to cover other needs.

Other external parties, such as TanCert, the Tanzania Organic Agriculture Movement (TOAM) and the district government, also play a vital role in creating a favourable environment in which the farmers and GFP can continue to collaborate, contributing to the development of the spices sector and its sustainability.

**Communication between partners**
The success of this partnership between spice producers and GFP is partly the result of their shared vision and the clear roles played by the various actors. The farmers' association, Juwakihuma, allows structured dialogue between them and GFP. They can negotiate common objectives and strategies, and discuss constraints or complaints at an early stage.

As the association and GFP mature, it is important that they agree on their respective roles in the value chain and in the extension work. External business development services and financial institutions also need to agree on their roles and on where to focus support. Disagreements and dissent and generally emerge when partners are not open about their agendas and what is at stake for them. It is important that all parties understand and respect each other's roles, interests and capacities.

**The way forward**
Investment in the spices sector can lead to profitable business ventures for smallholder farmers and SMEs alike, provided that minimum volumes are achieved in order to reduce unit costs. For GFP, it is necessary to strengthen the linkages with farmers and the farmers' association, through training on quality and farming as a business, and maintaining transparency on contract issues. At the same time, the company will have to develop its overall capacity for financial management, marketing and laboratory testing in order to reduce costs.

For more information, visit www.gfp.co.tz

## 7.2 Discussion

Trust, transparency and time are essential for establishing long-term relationships between actors both within a value chain and beyond. Financial institutions, service providers, government institutions and various other stakeholders, all affect the relationships between actors in a chain.

### Agreements

In this chapter, José Ruijter has reflected on the importance of trust in the relationship between producers and buyers. Golden Food Products (GFP) offers its producers supply contracts, but whenever they are in need of cash the farmers sell their spices to traders or middlemen, reducing the quantities available to GFP. This opportunistic behaviour is understandable but difficult to prevent. GFP invests a lot in proving its credibility to the farmers. It offers fair prices, support for social programmes and long-term trade relations. However, this approach is expensive and not all small businesses are able to invest so much time and money in building partnerships with their producers. In such cases, existing structures could be a starting point (such as involving traders in a partnership). It is important that there are rewards for keeping to an agreement, as well as sanctions in the event that one partner does not keep his word to support a relationship.

### Partners

Collaboration in spice chains requires that individual actors work together in partnership. It is important to select partners carefully. Selection can be based on rational criteria such as company output, financial performance and track record. But other factors are equally relevant, such as social performance and reputation, compliance with CSR guidelines and ILO standards, and whether they are involved in other partnerships. But perhaps the most important factor is compatibility, or 'like-mindedness'. In the case of GFP, for example, the Swiss company Organic Food Ingredients Sagl (OFI) was impressed by the owners. They shared a vision of how to do business and how they could work together to serve their common interests.

### Dialogue

For any partnership to work, dialogue is crucial. Dialogue promotes transparency, which in turn enhances trust. For GFP, a shared vision and clearly defined roles were also important. Dialogue is important vertically between actors in the chain, but also horizontally, such as between producers. In the case of GFP, farmers are better able to share their issues and wishes after discussions among themselves. Dialogue with actors outside the value chain is also important. Many value chains depend on financial institutions and other service providers. Disputes usually emerge when the partners are not open about their agendas and their interests. Understanding each other's roles and interests can help to prevent problems.

Selecting partners, building trusting relationships and developing a shared vision can take time, requiring all the partners to adopt a long-term perspective.

**A final note ...**
Some producers may be included in the partnership, while others are not. The drivers of the partnership should always reflect on what it means for those who are excluded.

Often the aim of partnerships with producers is to cut out the middlemen, but the stakeholders should always reflect on the implications of doing so. For example, processing companies may convince producers to work in partnership by paying more than local traders, although this might have the undesirable consequence of driving up local prices. Another result of cutting out the middlemen is that producers may become completely dependent on the partnership.

# 8 Towards sustainable trade

Bart de Steenhuijsen Piters, Royal Tropical Institute, Amsterdam, the Netherlands

## 8.1 A short synthesis

The global food system is linked to many of the challenges currently confronting humanity, such as climate change, food scarcity, natural resource degradation and poverty. To meet these challenges, the food system needs to change and become sustainable. One commodity that can make an important contribution in this area are spices, which are found in almost every food product. They are important from a poverty alleviation perspective because the majority of producers are poor smallholders. Moreover, the production of spices is regarded as potentially positive for the environment because spice crops can contribute to biodiversity.

The growing global demand for spicy and flavored food presents an opportunity for the spice sector. However, the challenges to supplying sustainable spices are considerable. Sustainable trade has strict requirements that are difficult for smallholders to comply with. As a result, smallholders are attracted to less stringent markets, or to crops that might yield a higher income. 'Dividends,' in the form of higher-value produce, will take time to develop through often complex value chains. Sustainability involves investments and a clear division of responsibilities.

Transformation of spice chains can be supported by standards and certification systems. Standards enable a company to analyse its own practices, together with that of its suppliers and employers, and standards force chain actors to comply with a norm. Currently, there are no specific standards for spices, and a process of standard-setting is required to address sector-specific characteristics. Such a process involves investigating how different actors in the spice chain perceive sustainability, setting actor-specific goals and designing action trajectories.

But standards alone will not move the sector towards sustainability. An essential ingredient is cooperation between chain actors and supporters. A transition to sustainability will require the involvement of producers, traders and retailers, but also private venture capital firms to provide investments, and financial institutions to provide trade capital. Moreover, public investment in education and training will be needed, as well as the creation of farmer organizations. It is a shared responsibility and demands careful alignment of activities along the chain. For such cooperation to succeed, long-term relations are needed that foster trust and transparency between actors. Building these relationships will require a significant investment of time and money.

Within such a complex web of actors, dedicated leadership is needed to coordinate and oversee the transition. Consumers can exert pressure on retailers by showing their willingness to pay more and by demanding sustainable products. However, they are not in a position to facilitate transformation. And the Dutch government is prepared to support change in the sector, but the drivers of change are the industry leaders. They are the ones in a position to take the first step and involve the rest of the industry.

The cases in this book demonstrate both the willingness of the spice sector to move towards sustainability, and the need for such a move. Nevertheless, many issues, some of them sector-specific, need to be addressed first. These vary from defining sustainability to bringing it into practice, from planting perennial trees to building lasting relationships based on trust.

## 8.2 Epilogue

In the Netherlands, most spice companies and traders are members of the Dutch Spice Trade Association, which could provide the critical mass necessary to transform the market. However, achieving a sustainable spice sector means going beyond value chains and involving actors from other sectors, financial institutions, NGOs and public research institutes. Knowledge of the process of producing spices, as well as of trade and retailing, needs to be linked to proposals for making spice value chains more sustainable. So far, few actors in these sectors have exchanged knowledge in these areas.

Several commodity sectors, including cocoa, tea and coffee, have already made progress in their quest for sustainability. Although the spice sector has its own peculiarities, it can learn from these other sectors. For example, the spice sector could learn how the cocoa sector standardized the quality of large volumes of produce: how did smallholders organize themselves to obtain this remarkable result? And the tea sector could provide insights into how it organized its extension services to smallholders. The spice sector could also learn how public-private cooperation has contributed to the success of other sectors.

Any recipe for sustainability in the spice sector should not remain a family secret but should serve to inspire other spice chains, sectors, producers and consumers. The Dutch Spice Trade Association can contribute significantly to furthering sustainability by creating a platform for collaboration and knowledge exchange. But dedicated assistance from other service providers is still needed to achieve the desired transformation. The Dutch Sustainable Trade Initiative could function as a broker and, with inputs from knowledge centres such as the Royal Tropical Institute, could support the sector on its journey towards a sustainable spice trade.

# 9 The way forward

Poul Wiertsema, Marco Eekelder, Nanto Prasetvo, Evert-Jan Verschuren;
Corporate Social Responsibility Working Group of the Dutch Trade
Spice Association, the Netherlands

It was a leap of faith for members of the Dutch Spice Trade Association (NVS) when they agreed to participate in a new working group on sustainability in the sector. An inspiring presentation from Bernedine Vos of MVO Netherlands had given association members food for thought: something clearly had to be done about corporate social responsibility in the spice sector, but what was not yet clear. Soon after that presentation, the NVS was approached by the NGO Both Ends and the consultancy company Crem. They signed a cooperation agreement to investigate opportunities for making the spice trade more sustainable. The NVS sustainability working group was happy to have a concrete project to work on, and the results were published in the report *'Can you pass me the sustainable pepper please? Quick facts about sustainable spices for European companies and producing countries'*. The report does not have all the answers, but it is a useful starting point for investigating what sustainability would mean for the spice sector.

Working with the Dutch Sustainable Trade Initiative (IDH) was the logical next step. In 2010, IDH and the Dutch Spice Association initiated a sustainability programme for spices and herbs. The programme will further look at what sustainability means for the spice sector: what is the current situation and how can it be improved? Issues such as preventing child labour and the contribution of spices to biodiversity will be part of the programme. Because the spice sector has a huge diversity of products, it is impossible to take into account all spices at this stage. Two spices, pepper and nutmeg, have therefore been chosen as a pilot.

In the coming period, through research and the exchange of experiences, we aim to set a sustainability standard for our sector. It will initially focus on international chains. The standard should be dynamic and applicable both to producing countries and consumer countries. The challenge will be to design and implement standards that are neither too low, requiring higher standards be set over time, nor too high, which would exclude many actors.

A successful initiative requires the support of the sector. On behalf of the working group, we would like to call on all stakeholders in the spice sector to support these initiatives and join the 'coalition of the willing'. The enormous participation of Dutch and foreign stakeholders in the *Sustainable Spice Conference: from tropical root to responsible food* (12 October 2010, Royal Tropical Institute) made clear that there is widespread interest in, and support for, sustainability. Now is the time to convert this support into action: contact the members of the working group or the secretary of the Dutch Spice Association to find out how.

On behalf of the Corporate Social Responsibility Working Group of the Dutch Trade Spice Association,

# About the authors

**Cleopa Ayo**
    General manager, Golden Food Products (GFP), Arusha, Tanzania
Mr Cleopa John Ayo is Managing Director of GFP Limited, a company that was established in 1998 by Mr Ayo and his wife, and is based in Arusha, Tanzania. Since 2006 GFP has set up an export business in organic spices from Usambara Mountains in Tanga Region.
    www.gfp.co.tz

**Mark Barnett**
    General director, Son Ha Spice, and managing director, Pacific Basin Partnership, Vietnam
Mark is an American who has been processing spices in Vietnam since 1994. His companies grow, process and export spices to industrial processors and retail packers in the European Union, Japan and the United States. A former wild and crazy guy, Mark is now married with children. He enjoys editing sustainability presentations, elaborate jokes and bad puns.
    www.pbpspice.com

**Patrick Barthelemy**
    Owner, Cassia Co-op, West Sumatra, Indonesia
Patrick, the founding member of Cassia Co-op, has worked in the spice industry for 18 years, focusing on cassia (Indonesian cinnamon) and vanilla. He started organic cassia projects in Sumatra as far back as 1995, and has set up profit-sharing systems to benefit growers and secure supply chains. He has marketed Indonesian cinnamon mainly in the United States and more recently in Europe. After years of working within conventional spice distribution channels, he finally created Cassia Co-op to implement his vision of this vision of full transparency and fair trade in the industry.
    www.cassia.coop

**Karin Bogaers**
    Coordinator, social compliance, product integrity, corporate responsibility, Royal Ahold, the Netherlands
Karin works for the corporate responsibility department of Royal Ahold, a major food retail company with operations in Europe and the United States. As the coordinator of social compliance, her responsibilities include the development

and enforcement of group policies in the fields of social compliance and sustainability. She is involved in setting industry-wide standards setting organizations in the fields of social compliance, working conditions and critical commodities.

    www.ahold.com

### Marije Boomsma
*Senior advisor, Royal Tropical Institute, Amsterdam, the Netherlands*

Marije Boomsma is a senior advisor on sustainable economic development at KIT, specialized in sustainable markets and value chains. Before joining KIT, Marije worked for several years as a market development consultant in South-East Asia. She is frequently invited to be a panel speaker or to lead debates.

    www.kit.nl

### Alex Bruijnis
*Ministry of Economic Affairs, Agriculture and Innovation (EL&I), the Netherlands*

Alex is policy advisor on sustainable commodities at the Ministry of Economic Affairs, Agriculture and Innovation (EL&I). His areas of expertise are environmental economics and organizational development.

    www.minlnv.nl

### John Fagan
*Director, Earth Open Source Institute*

John is the brains behind Earth Open Source Institute, a new organization that is in the process of creating a collaborative platform to support the development of 'open source'. The idea behind open source is that by exploiting the potential of networked collaboration, the whole world could become the social and environmental auditor of the food industry. Transparent supply chains and crowd-sourced 'ethical' evaluations of food products would be just the start.

    www.earthopensource.org

### Michael Gravina
*Founder and managing director, Elephant Pepper/Piri Piri Elefante, Mozambique*

Michael is the founder of Elephant Pepper, a chilli farming and processing company with operations across Southern Africa. The company promotes sustainable agricultural practices for strong and dynamic rural economies, and in which human and wildlife populations live in harmony. Michael is convinced that economic opportunities that tap into the entrepreneurial spirit of the world's poor are key to ending poverty. Rather than pity, what they want is access to opportunities such as those that Elephant Pepper, and a new venture, Piri Piri Elefante Mozambique, now provide.

    www.elephantpepper.com

### Marianne van Keep
*Director of procurement, Verstegen Spice & Sauces BV, and chair, Dutch Spice Trade Association, the Netherlands*

Marianne has worked for the last ten years as director of procurement at Verstegen. She is responsible for a wide range of products, of which spices have a strategic importance. Since 2005, Verstegen has been active in the field of corporate social responsibility (CSR) with respect to both the environment and fair trade. Strategy and implementation are partly the responsibility of the procurement department. Marianne has been chair of the Dutch Spice Trade Association since 2008, and is working to promote fair trade in the spice sector.

www.verstegen.nl

### Ellen Mangnus
*Junior advisor, Sustainable Economic Development, Royal Tropical Institute, Amsterdam, the Netherlands*

Ellen has worked at the Royal Tropical Institute since 2009. Her special interests include farmer organizations and access to markets.

www.kit.nl

### Bharat Maskai
*CEO, Laxmi Enterprises, Mumbai, India*

Bharat is the managing partner and CEO of Laxmi Enterprises, a leading exporter of spices (including steam-sterilized value-added spices) and spice blends to the world market. He is currently responsible for purchases, sales and production planning. Laxmi Enterprises pays fair prices and has established good links with farmers to ensure consistent high-quality supplies.

www.sbslaxmi.com

### Samash A. Nathu
*Director, Uganda Crop Industries Limited (UCIL), Kampala, Uganda*

Samash is the director of Uganda Crop Industries Limited (UCIL), a family-owned company that was founded in 1993. From its own fields, and with a network of small farmers across Uganda whom the company finances, trains and supports, the company produces high-value aromatic crops such as vanilla and cardamom for domestic, regional and international markets. The company also produces timber and sugar for local use, and encourages reforestation and sustainable farm practices.

www.ucil-uganda.com

### José Ruijter
*Programme officer, Catholic Organisation for Relief and Development Aid (Cordaid), the Netherlands*

www.cordaid.nl

### Bart de Steenhuijsen Piters
Area Leader, Sustainable Economic Development, Royal Tropical Institute, Amsterdam, the Netherlands

Bart holds a PhD in agro-system theory and is an advocate of diversity thinking. His current activities concentrate on value chain development and sustainable equity investments in developing countries.

www.kit.nl

### Sarah Simpson
Communications officer, Development Policy and Practice, Royal Tropical Institute, Amsterdam, the Netherlands

Sarah is a communications officer within the Royal Tropical Institute's main department for international development. Before joining KIT, Sarah worked as an editor and technical writer. She holds a BSc degree from the University of Toronto, Canada.

www.kit.nl

### Tjeerd de Vries
Ministry of Economic Affairs, Agriculture and Innovation (EL&I), the Netherlands

Tjeerd de Vries is a board member of the department for agribusiness and fisheries in the Ministry of Economic Affairs, Agriculture and Innovation. He is also a member of the board responsible for sustainable commodities, economy & strategy, agro-logistics, development cooperation and research.

www.minlnv.nl